Power Systems

Electrical power has been the technological foundation of industrial societies for many years. Although the systems designed to provide and apply electrical energy have reached a high degree of maturity, unforeseen problems are constantly encountered, necessitating the design of more efficient and reliable systems based on novel technologies. The book series Power Systems is aimed at providing detailed, accurate and sound technical information about these new developments in electrical power engineering. It includes topics on power generation, storage and transmission as well as electrical machines. The monographs and advanced textbooks in this series address researchers, lecturers, industrial engineers and senior students in electrical engineering.

Power Systems is indexed in Scopus

Ahmed Rachid • Aytac Goren • Victor Becerra •
Jovana Radulovic • Sourav Khanna

Solar Energy Engineering and Applications

Ahmed Rachid 🆔
UFR-Sciences
University of Picardie Jules Verne
Amiens, France

Aytac Goren 🆔
Mechanical Engineering
Dokuz Eylul University
Izmir, Türkiye

Victor Becerra 🆔
School of Energy and Electronic
Engineering
University of Portsmouth
Portsmouth, UK

Jovana Radulovic 🆔
School of Mechanical & Design
Engineering
University of Portsmouth
Portsmouth, UK

Sourav Khanna 🆔
School of Energy and Electronic
Engineering
University of Portsmouth
Portsmouth, UK

ISSN 1612-1287   ISSN 1860-4676   (electronic)
Power Systems
ISBN 978-3-031-20832-4   ISBN 978-3-031-20830-0   (eBook)
https://doi.org/10.1007/978-3-031-20830-0

This Springer imprint is published by the registered company Springer Nature Switzerland AG
The registered company address is: Gewerbestrasse 11, 6330 Cham, Switzerland

Preface

Solar energy is becoming a necessity not only for climate change issues but also for economic reasons because it has the cheapest energy production cost nowadays. This book gives a general and concise presentation of solar energy from a practical engineering perspective. Without covering detailed, in-depth physics but still giving a comprehensive, accessible, and intuitive introduction to the topic, this book focuses on presenting proven methods and tools for the design, implementation, and monitoring of solar energy systems, as well as associated auxiliary technologies. The book covers key aspects of solar energy, such as photovoltaic solar cells and systems, battery technologies, solar concentrators, and hybrid photovoltaic/thermal systems. Key application areas such as homes, buildings, solar farms, street lighting, vehicles, and pumping are discussed. The methods for connecting solar farms and other photovoltaic installations to power distribution systems are addressed in the context of the smart grids technologies available to facilitate such connections. The feasibility study of solar projects is also considered and illustrated by a practical example considering the technical, financial, and environmental aspects.

The chapters in this book are self-contained and may be read in any order by advanced students, practicing engineers, and researchers interested in solar energy applications, a multidisciplinary subject including physics, electrical engineering, thermal engineering, and power electronics.

Amiens, France
Izmir, Türkiye
Portsmouth, UK
Portsmouth, UK
Portsmouth, UK

Ahmed Rachid
Aytac Goren
Victor Becerra
Jovana Radulovic
Sourav Khanna

Acknowledgment

This work was carried out as part of the SOLARISE project of the Interreg 2 Seas program co-financed by the European Regional Development Fund sub-grant contract No. 2S04-004.

Contents

Chapter 1
Fundamentals of Solar Energy

1.1 Introduction to Solar Energy

Electromagnetic radiation emitted by the nearest star reaches the earth as *solar radiation*. Sunlight consists of visible and near visible regions. The *Visible region* is the region where the wavelength is between 0.39 and 0.74 μm. The infrared region has a wavelength smaller than 0.39 μm and the ultra-violet region's wavelength is greater than 0.74 μm.

Luminosity is the radiant power measure that is independent from the distance of the source and the observer. The luminosity of the Sun is 3.846×10^{26} [W/s] [10]. Solar radiation reaches the earth as direct, diffuse or reflected radiation. *Direct solar radiation* is the sunlight that directly reaches the surface. *Diffuse solar radiation* is the sunlight scattering through atmosphere whereas *reflected radiation* is the sunlight that reaches the surface via reflections from buildings or from other objects. The sum of these three types on a unit area is called as *global solar radiation*. *Solar irradiation* is the solar energy received by a 1 m^2 surface. Solar energy is reflected, absorbed and scattered before reaching the earth (Fig. 1.1).

When solar radiation passes through the atmosphere, some of it is absorbed or scattered. Clouds, air molecules, aerosols, and water vapor are the reasons. *Direct normal irradiance* (DNI) is the solar radiation that reaches directly the earth's surface. *Diffuse horizontal irradiance* (DHI) is the terrestrial irradiance received by a horizontal surface that has been scattered or diffused by the atmosphere. The total amount of shortwave terrestrial irradiance received by a surface horizontal to the ground is called as *global horizontal irradiance (GHI)*.

The value of solar radiation at mean earth-sun distance at the top of the atmosphere is referred to as the *Solar Constant* which is 1367 W/m^2. On a surface on earth in a clear day, at noon, the direct beam radiation can be approximately 1000 W/m^2. Location, season, time of the day and weather conditions are the main factors that affect the harvesting of solar energy [14, 15, 17].

© The Author(s), under exclusive license to Springer Nature Switzerland AG 2023
A. Rachid et al., *Solar Energy Engineering and Applications*, Power Systems,
https://doi.org/10.1007/978-3-031-20830-0_1

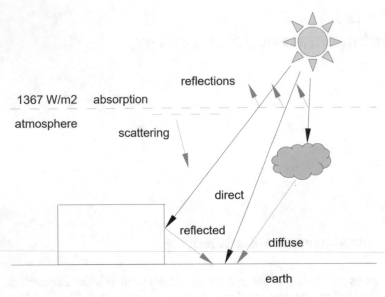

Fig. 1.1 Solar radiation on earth surface

Solar energy systems are the systems that use solar energy as a thermal source or generate electricity directly. Solar energy systems which use solar energy to generate electricity use the *photovoltaic effect*. Antoine-César and Alexandre-Edmond Becquerel have first observed the photovoltaic effect in year 1839. They noticed electrochemical effects produced by light in electrolytic solutions. In 1905, Albert Einstein explained the photoelectric effect caused by photons [4]. In 1932, cadmium selenide and in 1954 silicon-based cells were discovered to have photovoltaic effects. For the systems use solar energy as a thermal source, once again the energy comes from the photons. This is because photon kinetic energy is transferred to the object and it heats up the object. The Stefan-Boltzmann law explains the relationship between an object's temperature and the amount of radiation that it emits.

The amount of energy that can be generated by photovoltaic effect is related to the amount of solar radiation and how long it is received. Thus, the measurements to analyze the energy can be generated by solar energy systems are based on these two parameters.

1.2   Measurement of Solar Radiation

The earth orbits the sun and rotates around its own axis. The radius of the earth's orbit around the sun changes from 147 million kilometers to 152 million kilometers. The axis of the Earth is inclined with respect to its orbital plane at an angle of 23.45°.

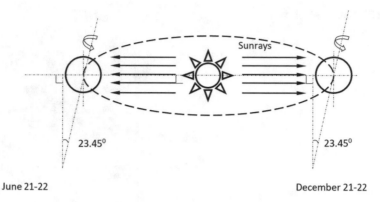

Fig. 1.2 Earth motion around sun and around its own axis

This geometry, and distance changes from the sun forms the seasons and a major change in solar radiation received by locations (Fig. 1.2).

Terrestrial radiation is a term used to describe infrared radiation emitted. German Physicist Wilhelm Wien defined the product of absolute temperature and wavelength as a constant. This makes it understandable why the solar radiation is called as *shortwave radiation* whereas terrestrial radiation is called as longwave radiation. Approximate ranges of defined regions are:

- Ultraviolet: 0.20–0.39 [μm]
- Visible: 0.39–0.78 [μm]
- Near-Infrared: 0.78–4.00 [μm]
- Infrared: 4.00–100.00 [μm]

Approximately 99% of solar radiation at the earth's surface is contained in the region from 300 nm to 3.0 μm whereas most of terrestrial radiation is contained in the region from 3.5 μm to 50 μm.

Campbell – Stokes sunshine recorder is a basic but an effective system to record hours of bright sunshine. It was invented by Francis Campbell in 1853 and then modified by Gabriel Stokes in 1879. It is composed of a 10 cm crystal sphere mounted to make a trace on the record material with focusing the sun rays. It has no moving parts. Since the focus location changes in day hours, the hours of bright sunshine are recorded on the record card. In Fig. 1.3, a basic drawing of a Campbell-Stokes Recorder can be seen.

1.2.1   What Is Net Radiation Measurement?

Net radiation (R_n) is the sum of incoming and outgoing radiation at the surface (Eq. 1.1). In Eq. 1.1, α is the albedo, R_s is the sum of direct and diffuse radiations,

Fig. 1.3 Campbell-Stokes
Sunshine Recorder. [5]

R_{il} is the incoming longwave, and R_{ol} is the outgoing longwave radiation. Albedo is the ratio of the incoming shortwave divided by the reflected shortwave.

$$R_n = (1 - \alpha)\, R_s + R_{il} - R_{ol} \tag{1.1}$$

For net radiation; shortwave, reflected shortwave, incoming and outgoing longwave need to be measured.

1.2.2 Direct, Diffuse and Global Measurements

Solar energy potential must be considered before installations of solar energy systems to the location. Solar energy potential can be analyzed using measurements and measurement based calculations. It is ideal to have at least 1 year of measurements.

Specification and classification of instruments for measuring hemispherical solar and direct solar radiation is defined in standard ISO 9060:2018. This document establishes a classification and specification of instruments for the measurement of hemispherical solar and direct solar radiation integrated over the spectral range from approximately 0.3 [μm] to about 3 [μm] to 4 [μm].

1.3 Instruments for Measuring Solar Radiation

Shortwave radiation defines the solar radiation in the visible, near-ultraviolet, and near-infrared spectra.

The *pyrheliometer* measures the solar radiation at *normal* incidence (G_n), so direct irradiance is best measured by a pyrheliometer. It is designed to measure the instantaneous value of the direct beam irradiance so, the structure is a tube with a photosensitive component at the end (Fig. 1.4a). The acceptance angle of the tube is 5° and inside the tube is coated black to absorb the light coming to the inner side of

(a)   (b)

Fig. 1.4 (**a**) Pyrheliometer (Wikipedia) and (**b**) pyranometer (DECharge 3 Research Station, Dokuz Eylül University)

the tube walls. Since it measures the direct light, it should be positioned pointing to the sun. For this reason it is common to combine the pyrheliometer with a two-axis sun tracker.

The *pyranometer* (Fig. 1.4b) measures the global irradiance. Using a shading disk or a ring, the diffuse irradiance can be measured by a pyranometer (Fig. 1.5). Incoming and outgoing longwave radiations are measured by precision infrared radiometer, pyrgeometers. In order to analyze the solar energy potential, sunshine duration is needed. The World Meteorological Organization (WMO) defines the sunshine duration as the time during which the direct solar radiation exceeds the level of 120 Watts per square meter.

A pyranometer is an instrument to measure the global solar irradiance coming to a surface (see Fig. 1.5). It receives the light in an angle of 180°. There are two types of pyranometers, thermoelectric type and photodetector type. In thermoelectric pyranometers, a thermopile is used as a sensor. Two concentric hemispherical glass cover the black colored sensor. These two domes shield the sensor from thermal convection and protect it against outdoor weather conditions. Silica gel inside the dome absorbs vapor. A white disk is used to limit the acceptance angle to 180°. The temperature difference between the layers gives a voltage output. The devices in this group are capable of detecting a wide range of atmospheric solar radiation spectrum (300–4000 nm). On the other hand, a photodetector pyranometer uses semiconductors to measure the irradiance reaching the surface. These type of pyranometers are not able to detect the full atmospheric solar radiation spectrum (app. 350–1100 nm). Since they use semi-conductor components, the response has a peak between 900 and 1100 nm. The output of a pyranometer is generally connected to a data logger to analyze the solar energy generation at the location throughout the year. Two common representations in solar radiation measurements are monthly average daily total radiation on a horizontal surface and hourly total radiation on a horizontal surface.

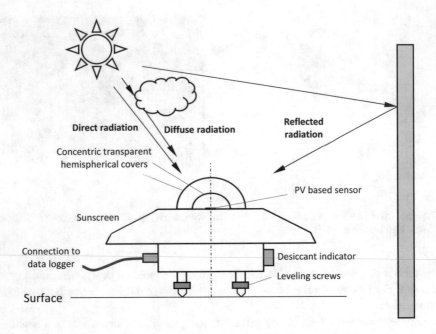

(a) Sensing global irradiance

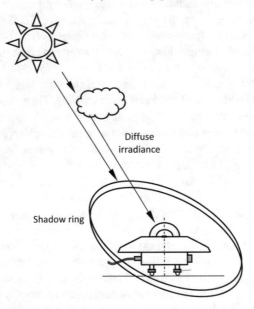

(b) Sensing diffuse irradiance with a shadow ring

Fig. 1.5 Pyranometer

(a) Solar power meter   (b) Lux meter

Fig. 1.6 Handheld solar power meter that uses silicon photovoltaic detector and a luxmeter

In order to measure the diffuse solar irradiance, a shadow ring can be used as illustrated in Fig. 1.5b. Since the daily sun elevation angle changes day by day, the angle of the shadow ring needs to be changed in this use [12]. As an alternative, a shadow disk, which tracks the sun elevation angle automatically can be also used for this purpose [11].

In mobile applications, like solar cars, handheld solar power meters are very useful to sense incident solar energy and fault detection (Fig. 1.6a). Indoor light levels (200–1000 lx) are very low to generate electricity compared to outdoor light levels (~10^5 lx). Nevertheless, there are devices that use solar cells indoors. In indoor measurements, lux meters (Fig. 1.6b) can be used with related calibration [16].

Handheld solar power meters are useful devices to quickly measure instantaneous solar irradiance or to measure the energy generation in mobile conditions. Table 1.1 summarizes which device and method can be used for which measurements.

1.4   Photovoltaics Cell Types

1.4.1   Silicon PV Cells (Monocrystalline, Multi or Polycrystalline and Amorphous Silicon Cells)

Silicon crystal (c-Si) type of solar cells are the most common PV cells in the market. According to The National Renewable Energy Laboratory (NREL) of the United States Department of Energy, the efficiencies of non-concentrated mono crystalline Si cells increased from 14% to 26.1% in the years between 1977 and 2022 [1].

Table 1.1 Which device for which solar energy measurements

Measurement	Device and method
Global irradiance	Pyranometer
Direct solar radiation	Pyrheliometer on an adjustable plane
Diffuse solar radiation	Pyranometer with a shadow ring or shadow disk with a tracker
Reflected solar radiation	Pyranometer
Solar radiation on a tilted surface	Pyranometer
Ratio of the reflected solar radiation by a surface to the total solar radiation (albedo).	Two pyranometers or systems having two pyranometer sensors
Solar vehicle energy generation	Handheld solar power meters or independent cell measurement setup
Indoor light measurement	Lux meter with related calibration

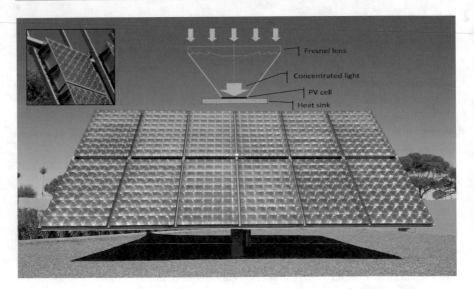

Fig. 1.7 A sun tracker with CPV and illustration of concentrating light on PV cell. (Photo: A.Gören)

In recent years, multicrystalline Si cells reached to 23.3% efficiency. They are not transparent but creating holes in cells in manufacturing process provides limited transparency for architectural applications. The high production of silicon PV cells prompts researchers to consider life-cycle assessment of such cells [3].

The *Concentrator* is the part of the module that concentrates the sunlight ten to one thousand times on the PV cell. Fresnel lenses and mirrors are used for this purpose [19]. Concentrating photovoltaics (CPV) cells are need to be cooled with heatsinks since the concentration of sunlight heats up the cell more than non-concentrated ones. A photo of a concentrator PV can be seen in Fig. 1.7.

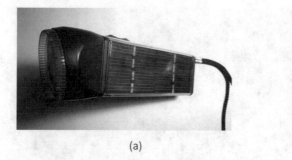 

(a)   (b)

Fig. 1.8 Solar flashlights using multi crystalline (**a**) and mono crystalline (**b**) Silicon PV cells. (Photo: A.Gören)

The theoretical efficiency of a simple Si cell is defined with Shockley–Queisser limit. That is 30% at 1.1 eV. Theoretical calculations show that a two-layer cell can reach 42% efficiency whereas a three-layer cell can reach upto 49%. A theoretical infinity-layer cell without any concentrator can reach up to 68% under sunlight.

Today, most of the solar arrays on roofs of the houses are composed of mono crystalline or multi crystalline PV modules. c-Si cells are implemented in different outdoor devices as well. It is very common to see c-Si cells on solar garden lights or flashlights like in Fig. 1.8.

On the other hand, almost all indoor solar energized devises use thin film solar cells. Since amorphous silicon (a-Si) cells perform better at low light levels, usually amorphous silicon solar cells are used in solar energy harvesting in indoor use. Mono or multi crystalline Silicon cells have a thickness of 200–300 microns whereas the amorphous ones have a thickness of 2–6 microns. Solar calculators, solar wristwatches, handheld solar devices, wireless sensors, garden lights use a-Si PV cells. Implementation of solar energy harvesting in wristwatches started in early 70s. In addition to very interesting design LED timepieces like Synchronar 2100 or Nepro Alfatronic watches; Seiko (A156-5000), Citizen (Crystron) and other known companies also produced digital or analog solar watches in these period. Although the first ones were emphasized the solar power with their designs, over time, light powered timepieces' designs became more natural. Nowadays, with different types of solar cells and implementation of PV cells in layers of design, it is not easy to tell if a watch is solar powered if it is not mentioned (Fig. 1.9).

The characteristic open circuit voltage of an a-Si cell is 0.89 V and its short circuit current is 14.8 mA/cm^2 under AM1.5, 1000 W/m^2 and 25 °C (Panasonic, amorton). A triple junction a-Si cell absorbs and use the blue, green and red light wavelengths respectively, which makes it possible to produce energy efficiently through a wide range light spectrum.

One of the interesting implementations was a solar energy powered cell phone manufactured by Samsung in the year 2009 (Fig. 1.10). On the backside of the device, there were solar cells that can make a ten minute talk after leaving the phone one hour under the sun.

Fig. 1.9 Analog and digital solar watches with a-Si PV cells. (Photos: A.Gören)

Fig. 1.10 Solar cell phone with amorphous Silicon PV cell that is produced in year 2009. (Photos: A.Gören)

There are also hybrid a-Si cells. These are called as heterojunction with intrinsic thin layer (HIT, SANYO; after 2010 Panasonic-Sanyo). In this type of cells, the mono n type c-Si layer is covered with n-type a-Si on one side, and n-type a-Si layer on the other side. This structure increases the cell efficiency.

One of the interesting and increasingly popular research area in PV cells is transparent solar cells. There are different types of transparent PV (TPV) cells. Luminescent solar concentrator technology and visible light partial absorption are the two main methods to achieve transparency [8].

1.4.2 Thin Film Copper Indium Gallium (di)Selenide (CIGS), Cadmium Telluride (CdTe) and $Cu_2ZnSn(S/Se)_4$ (CZTS) Cells

The most common thin film cell types are a-Si, Copper Indium Gallium (di) Selenide CGIS and Cadmium Telluride types. Cadmium telluride is the most cost effective among other thin film technologies and it has the lowest carbon foot print compared to the other technologies. Moreover, it offers the shortest payback period (less than 1 year). The efficiency of the CIGS cell has increased to over 23% in recent years. These type of cells perform better in higher temperatures, low light levels and shaded conditions than Si cells. Nevertheless, it should not be forgotten that cadmium is a toxic material and tellurium supply is not easy. N-doping process of CdTe is either with Al, Ga, In or with Fe, Cl, Br, I; p-doping can be with copper, silver, gold or with N, P, As, Sb.

1.4.3 Gallium Arsenide (GaAs) PV Cells

The single crystal GaAs single junction solar cell can reach efficiency close to 28% and the thin film GaAs cell can reach more than 29% efficiency. If the single junction GaAs cell is used with a concentrator, the efficiency can reach more than 30%. The most efficient PV cell is a multiple junction (four or more) gallium arsenide cell with concentrator and has an efficiency of more than 47% [1].

Multiple junction GaAs PV cells or thin film GaAs cells are used on satellites, solar cars or in research (Figs. 1.11 and 1.12). They are very expensive. Comparing to Si cells in the market, multiple junction GaAs cells' prices are twenty or more times more than efficient mono c-Si cells.

1.4.4 Perovskite PV Cells

Perovskite structure defines the structure similar to the inorganic mineral compound crystal, namely calcium titanium oxide ($CaTiO_3$, calcium titanate) structure. It is named after Russian mineralogist, Lev Alekseyevich von Perovski (1792–1856).

The perovskite crystal can be represented as ABX_3. It is composed of X anion, A and B cations. The structure of the perovskite can be seen in Fig. 1.13.

In Perovskite solar cells, common layers are the transparent layer and the back electrode layer, charge transport layers behind electrode layers and the perovskite layer. With more focus in research after 2012, Perovskite cell efficiencies reached from 10% to over 25% in 2020. The reason is the discovery of the thin stable perovskite layer. Perovskite solar cells can be manufactured in a simpler way compared to conventional silicon cells and this can provide cheaper solar cell

Fig. 1.11 GaAs and back
contact non-reflective layer
coated mono c-Si cells on the
same surface. (Photo:
A.Gören)

Fig. 1.12 Stanford University (US, single junction thin film GaAs), Antakari Solar Car (Chile,
monocrystalline Si), Solaris 10 Solar Car (Türkiye, monocrystalline Si). (Photo: A.Gören)

Fig. 1.13 Perovskite
structure

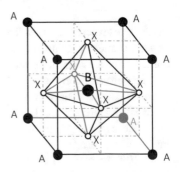

production [18]. On the other hand, the perovskite layer can be combined with silicon solar cells to provide silicon cells that use the blue light region and achieve higher efficiencies than the theoretical efficiency of simple Si cell (Shockley–Queisser limit, 30% @1.1 eV, 33% @1.34 eV). Perovslite/Si tandem cells reached more than 29% efficiency in 2022.

1.4.5 Organic PV Cells

Organic PV cells (OPVC) have a great importance in technological research [2, 7, 13]. They are known as polymer solar cells and consist of very light, flexible structures. One common electrode material of organic PV cells is indium tin oxide (ITO). Graphene, carbon nanotube and metal nanostructures are other alternative materials for electrodes. A single layer polymer PV cell is composed of ITO or metal on one side, aluminum, magnesium or calcium on the other side and organic electronic material between these two layers.

The types of carriers in polymers are polarons, excitons and solitons. When a charge carrier moves in the medium, the induced polarization follows its motion. The carrier together with the induced polarization is considered as polaron. An exciton is a bound state of an electron and an electron hole. Excitons and polarons play an important role in the electronic and optical properties of organic polymer PV cells. Light is absorbed in donor and stable exciton is generated. Diffusion of exciton provides polaron production. Free e^- and h^+ are generated from polaron pairs. Thus, current is generated.

OPVCs can be transparent, printed on a surface and potentially cost-effective. Types of OPVCs are bulk heterojunction thin-film solar cells, organic tandem solar cells, organic dye-sensitized solar cells.

OPVC efficiencies reached more than 18% in 2022. Considering that the efficiency of an OPVC is below 4% in year 2004, it can be said that one of the foci of the research in PV technologies is OPVCs. On the other hand, recent efficiencies of dye-sensitized cells are over 13% and tandem organic PV cell efficiencies are approximately 14% in 2022.

1.4.6 Thermophotovoltaic Cells

Thermophotovotaic (TPV) cells convert infrared light wavelength (up to 2.5–3.5 μms) into electricity using photovoltaic effect. The two parts of a TPV cell are a PV diode cell and a thermal emitter. TPV cells are based on Germanium (Ge), Silicon (Si), Indium gallium arsenide (InGaAs)/Indium phosphide (InP), Gallium antimonide (GaSb), Indium gallium arsenide antimonide (InGaAsSb), Aluminum gallium arsenide antimonide (AlGaAsSb), Indium arsenide antimonide phosphide (InAsSbP) and Indium arsenide (InAs) [6]. It is common to see these materials in photo sensors. Since they use IR, they operate at high temperatures (900–2400 °C). This condition limits the implementations to spacecraft applications, research focusing on continuous electricity generation or energy storage. Combining TPVs with a burning process is common. As research implementation, a TPV powered automobile called "Viking 29" was built by Western Washington University [9]. In this series hybrid electric vehicle, GaSb TPV cells were used to charge its 8.19 kWh NiCd battery pack. The silicon carbide emitter is heated to 1700 °C using natural gas. The car has a 100 hp DC brushless motor and the body has a polymer composite monocoque body.

References

1. Best Research-Cell Efficiency Chart (2022), NREL, Retrieved from https://www.nrel.gov/pv/cell-efficiency.html.
2. Choy, W.C.H. (2013). Organic Solar Cells, Materials and Device Physics, Springer London.
3. Dubey, S., Jadhav, N.Y., Zakirova, B.(2013), Socio-Economic and Environmental Impacts of Silicon Based Photovoltaic (PV) Technologies, Energy Procedia, Volume 33, 2013, pp. 322–334, https://doi.org/10.1016/j.egypro.2013.05.073.
4. Einstein, A., (1905). Über einen die Erzeugung und Verwandlung des Lichtes betreffenden heuristischen Gesichtspunkt, Annalen der Physik, 17(6), 132–148.
5. Encyclopædia Britannica (1911).
6. Gamel, M.M.A., Lee, H.J., Rashid, W.E.S.W.A., Ker, P.J., Yau, L.K., Hannan, M.A., Jamaludin, M.Z. A, Review on Thermophotovoltaic Cell and Its Applications in Energy Conversion: Issues and Recommendations. Materials, 2021, 14, 4944, https://doi.org/10.3390/ma14174944.
7. Głowacki, E.D., Sariciftci, N.S., Tang, C.W. (2012). Organic Solar Cells. In: Meyers R.A. (eds) Encyclopedia of Sustainability Science and Technology. Springer, New York, NY. https://doi.org/10.1007/978-1-4419-0851-3_466.
8. Lee, K., Um, H.-D., Choi, D., Park, J., Kim, N., Kim, H., Seo, K.(2020), The Development of Transparent Photovoltaics, Cell Reports Physical Science, 1(8), 2020, 100143, https://doi.org/10.1016/j.xcrp.2020.100143.
9. Morrison, O., Seal, M., West, E., Connelly, W. (1999), Use of a Thermophotovoltaic Generator in a Hybrid Electric Vehicle, AIP Conference Proceedings 460, 488 (1999); https://doi.org/10.1063/1.57831
10. New generation photovoltaics: A Guide to Design and Implementation Updated to Third Generation Technology, Fabio Andreolli, 2014.
11. Paulescu, M., Paulescu, E., Gravila, P., Badescu, V. (2013). Solar Radiation Measurements. In: Weather Modeling and Forecasting of PV Systems Operation. Green Energy and Technology. Springer, London. https://doi.org/10.1007/978-1-4471-4649-0_2
12. Sengupta, M., Habte, A., Kurtz, S., Dobos, A., Wilbert, S., Lorenz, E., Stoffel, T., Renné, D., Gueymard, C., Myers, D., Wilcox, S., Blanc, P., Perez, R. (2015). Best Practices Handbook for the Collection and Use of Solar Resource Data for Solar Energy Applications, Technical Report, NREL/TP-5D00-63112, February 2015, Retrieved from https://www.nrel.gov/docs/fy15osti/63112.pdf.
13. Tress, W. (2014). Organic Solar Cells, Springer International Publishing Switzerland.
14. Ulgen, K. and Hepbasli, A. (2004), Solar radiation models. Part 2: Comparison and developing new models, Energy Sources, 26(5), pp.521–530.
15. Unver, T., Goren, A.(2019), Development of seasonal solar radiation estimation models for Dokuz Eylul University campus area for controller of a two-axis solar tracker, Inter-

national Journal of Global Warming, 2019 , 19(1/2), pp.193–201, https://doi.org/10.1504/IJGW.2019.101781.

16. Venkateswararao, A., Ho, J.K.W., So, S.K., Liu, S.-W., Wong, K.-T., (2020), Device characteristics and material developments of indoor photovoltaic devices, Materials Science and Engineering: R: Reports, 139, 2020, 100517, ISSN 0927-796X, https://doi.org/10.1016/j.mser.2019.100517.

17. Wong, L.T., Chow, W.K. (2001), Solar radiation model, Applied Energy, 69 (3), 2001, pp. 191–224, https://doi.org/10.1016/S0306-2619(01)00012-5.

18. Wu, T., Qin, Z., Wang, Y. et al. (2021), The Main Progress of Perovskite Solar Cells in 2020–2021. Nano-Micro Lett. 13, 152 (2021). https://doi.org/10.1007/s40820-021-00672-w

19. Xie, W.T., Dai, Y.J., Wang, R.Z., Sumathy, K. (2011), Concentrated solar energy applications using Fresnel lenses: A review, Renewable and Sustainable Energy Reviews, 15 (6), 2011, pp. 2588–2606, https://doi.org/10.1016/j.rser.2011.03.031.

Chapter 2
Photovoltaic Cells and Systems

2.1 Fundamental Principles

2.1.1 Introduction to photovoltaic cells

The photovoltaic effect is the generation of electricity when light hits some materials. In 1839, Antoine-César and Alexandre-Edmond Becquerel were the first persons to observe electrochemical effects produced by light in electrolytic solutions [1, 2]. W. Grylls Adams discovered that selenium generates electricity when exposed to light in 1876. In 1883, C. Edgar Fritts made the first solar cell which consists of selenium covered by gold. Its efficiency was less than 1%. In 1932, Audobert and Stora discovered the photovoltaic effect of cadmium selenide (CdSe). But, the big step in PV cell research was the discovery of silicon cells in 1954 at Bell Labs. The efficiency of the first silicon cell was 6%, which was impressive. Today, silicon cells are very common in the market and some have efficiencies higher than 27%.

The photovoltaic cell is generally a constant current source which is directly proportional to the solar radiation falling on the cell. The equivalent electrical circuit of a solar cell consists of three functional layers. These are n-type layer, p-type layer and depletion layers. The depletion layer is the middle layer and the one connects both pole layers using the photon energy. The n-type layer is the layer which has free electrons and the p-type layer is the one which has holes for electrons. Photon energy converts the semi-conductor structure into a conductor and also converts light into electricity at the same time.

Silicon has four valence electrons. When silicon makes a bond with a group 15 element phosphorus (P), phosphorus donates one electron and forms the n-type layer. On the other hand, if silicon makes a bond with a boron atom, boron accepts one electron and forms the p-type layer. When solar radiation reaches the n-type layer of the PV cell, some photons hit the n-type layer and excite the electrons in the n-type layer and form the current flow (Fig. 2.1). For this reason, when the

A. Rachid et al., *Solar Energy Engineering and Applications*, Power Systems, https://doi.org/10.1007/978-3-031-20830-0_2

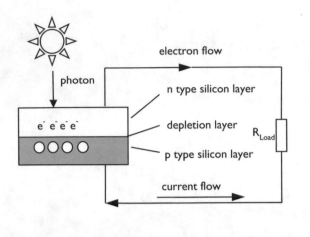

Fig. 2.1 Silicon photovoltaic cell layers

surface area of the cell increases, the current increases. The upper side of the PV cell is negative whereas the lower side is positive. The typical voltage of a Si PV cell is around 0.58 V.

2.1.2 Manufacturing of a Silicon PV Cell

Silicon cells are most common cells in the market and in research. A poly crystal silicon cell is formed with many crystals whereas the mono silicon PV cell is formed using one seed Silicon. Silicon has the atomic number 14 and four bonding electrons. In a 'single' bond, the silicon atom contributes just one electron to the bond; it can therefore form single bonds with four neighboring atoms. After forming the Silicon crystal structure which is very pure, n-Type and p-Type layers are formed by a doping process. As an example, phosphorus can be used for forming the n-type layer and Boron atom can be used for the p-type layer. It can be easily noticed if a PV cell is produced using multi crystal or mono crystal structure by looking to the surface of the PV cell.

Silicon photovoltaic cell manufacturing starts with growing the Silicon Crystal in a furnace (Fig. 2.2a). Today, the crystals can be grown to 200–300 mm diameter and 1–2 m length. By cutting the grown Si crystal at a thickness of 200–350 um, thin wafers (leaves) on which solar cells will be made are produced (Fig. 2.2b). After surface cleaning which can be either chemical or mechanical, the silicon wafer becomes ready for the diffusion process. Doping is adding impurities to a Si lattice. The aim is to increase the number of carriers. Pouring phosphoric acid on the boron-doped p-type wafer and exposing it to high temperature (800–1000 °C) allows the

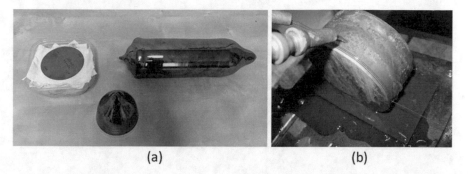

(a) (b)

Fig. 2.2 (a) Silicon crystal, (b) wafer cutting processes. (Eylem Project, EMUM & ACRL Lab, Dokuz Eylul University, 2014)

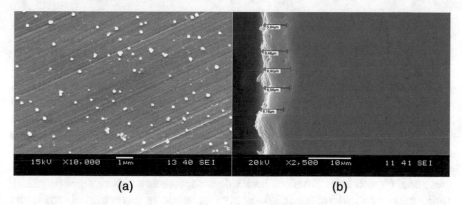

(a) (b)

Fig. 2.3 (a) Boron doped Si wafer SEM image, (b) Phosphorus-doped Si wafer on P-type wafer. (Eylem Project, EMUM & ACRL Lab, DEU, 2014)

phosphorus to diffuse into the wafer. As a result, an n-type joint is formed on the upper surface of the wafer.

Next step is Plasma Enhanced Chemical Vapor Deposition (PECVD). PECVD ensures that current collector paths are coated on the silicon formed by p-n junctions, and contacts are taken from the lower and upper parts of the photovoltaic cell. In addition, all kinds of surface coatings such as non-reflective surfaces can be made with PECVD. For example, making a non-reflective surface by coating TiO_2 is a very common method. The coated current collectors are heated and precipitated on the crystal surface and better contact is ensured. After production, cells are tested for their efficiencies and characterization. Figure 2.3a, b show the SEM images of the n-type and p-type layers which Boron and Phosphorus are doped.

In order to form PV modules from PV cells, the main processes are electrically combining cells and encapsulation. *Lamination* is the process of packing the PV cells in layers for mechanical protection. The main reason of lamination is to keep the efficiency of PV module longer. Most of the PV modules in market provide 25 years of life time or longer. EN 50513:2009 defines the data sheet and product

Fig. 2.4 PV module layers

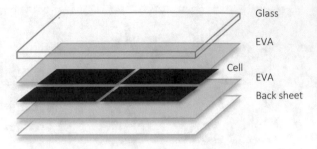

information for crystalline silicon wafers for solar cell manufacturing. EN 50461 defines the datasheet information and product data for crystalline silicon solar cells.

2.1.3 Lamination Materials and Techniques

Lamination is combining layers to form a structure which has the properties of the layers. The purpose of PV lamination is to protect PV cells with different technological materials to make them continue their function. Encapsulation means covering a functional layer with layers. The difference between encapsulation and lamination is that in lamination the layers are thicker.

There are at least five layers in lamination of PV cells in market. These layers are, glass, PV cell layer, front and back side encapsulants and back sheet (Fig. 2.4). Using ethylene vinyl acetate (EVA) as encapsulant is very common. These layers are for well protection of cells. Photovoltaic cells are very fragile and thin. The backsheet layer's main purpose is to increase the strength of the structure with a glass layer. In order not to filter any region of the effective light spectrum which is needed by the cell, encapsulation materials on the front side of the cell are very important. In solar car manufacturing, there is research on the lamination of the solar cells with polymer composites. Polymer composites are composed of two components: The matrix and the fibers. Glass fibers are suitable to strengthen the structure and do not affect optical transparency. Epoxy, which is the matrix part of the composite structure, should be selected carefully to avoid filtering the effective light spectrum and loosing the optical transparency in the number of years defined in the standards.

For implementation of polymer composites to PV cell lamination, vacuum assisted resin transfer molding technique is suitable. This technique applies vacuum to transfer resin from a tank to the vacuum isolated production ready structure. Curing is performed under vacuum. Polymer composite implemented PV modules are very light and strong. If we compare weights of a glass laminated PV module and a polymer composite laminated one, the weight of the polymer composite laminated PV module can be just 5% of the glass laminated one.

2.2 Ideal & One Diode Models

The equivalent circuit of a PV cell can be simply modeled as a current source in parallel with a resistor and a diode those are connected in series with another resistor. The output of the current source is directly proportional with the solar radiation falling on the cell. The open circuit voltage of the cell is quite different from the cell that is connected with the load which is shaped with R_s resistor. Besides, the temperature effects the output of the cell as well. On the other hand, the mathematical model of the cell is often defined with the Shockley Diode Equation. In Table 2.1, one and two diode models of a PV cell can be seen.

Photovoltaic cells are current sources where the current generated by them changes with the solar irradiation (solar radiation received per unit of area). If we look at Fig. 2.5a we can see that the current is directly proportional to solar radiation received by the surface of the cell. And in Fig. 2.5b, it can be seen that the increase in temperature decreases the open circuit voltage of the cell and increases the current it can generate. However, since cell is considered to be used at its maximum power point, the energy it can generate decreases with the increase in temperature.

Cells are connected to form PV modules; PV modules are connected to form solar arrays. Serial connection of cells increases the open circuit voltage whereas

Table 2.1 Common mathematical models of PV cell

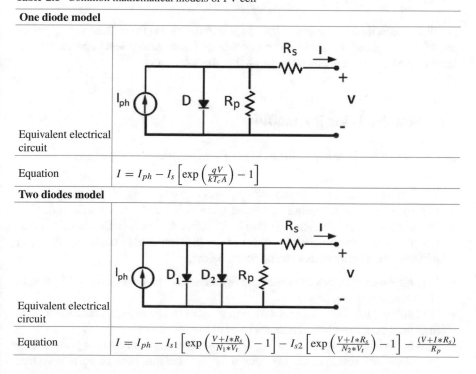

One diode model	
Equivalent electrical circuit	
Equation	$I = I_{ph} - I_s \left[\exp\left(\frac{qV}{kT_cA} \right) - 1 \right]$
Two diodes model	
Equivalent electrical circuit	
Equation	$I = I_{ph} - I_{s1} \left[\exp\left(\frac{V+I*R_s}{N_1*V_t} \right) - 1 \right] - I_{s2} \left[\exp\left(\frac{V+I*R_s}{N_2*V_t} \right) - 1 \right] - \frac{(V+I*R_s)}{R_p}$

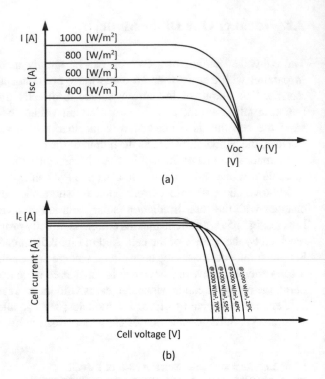

Fig. 2.5 (**a**) Generation of electricity of a PV cell under different solar irradiation (**b**) Cell voltages in different cell temperatures

parallel connection of modules increases current as in Table 2.2. The types of module connections are decided considering the input voltage level and the current level of the units which will be connected to the solar arrays.

2.3 Standards for PV modules

2.3.1 Testing Photovoltaics After Production

After production, photovoltaic modules are tested to determine if they are qualified. The International Electrotechnical Commission (IEC) and The American Society for Testing and Materials (ASTM) standards define design qualifications, describe type approval and test conditions. A list of the IEC standards directly or indirectly related with the photovoltaics can be found below;

- **IEC 60068-2:2022** is an international standard for the environmental testing of electrotechnical products.
- **IEC 60891:2021** is the standard for temperature and irradiance corrections to the measured current-voltage characteristics of photovoltaics.
- **IEC 60904-1:2020** is the most important standard for solar cells or photovoltaic modules since it describes procedures for the measurement of current-voltage

Table 2.2 PV module connections

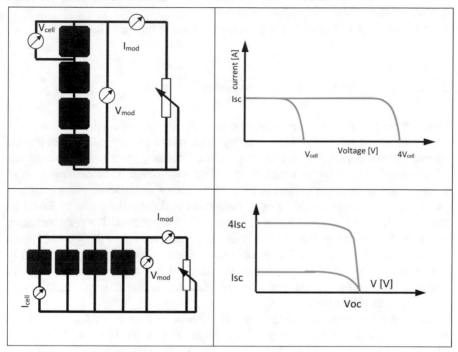

characteristics (I-V curves) of photovoltaic devices in natural or simulated sunlight.

- **IEC 61215** is suitable for terrestrial PV modules for long-term operation in general open-air climates. It indicates the requirements for the design approval, the type approval and the qualification process.
- **IEC 61701:2020** defines salt mist corrosion testing of PV modules.
- **IEC 61730** is a global standard for PV module safety qualification.
- **IEC 61829:2015** defines on-site measurement of current-voltage characteristics of PV arrays.
- **IEC 62108:2016 RLV:** defines design qualification and type approval of concentrator photovoltaic modules and assemblies.
- **IEC 62548:2016** defines design requirements for photovoltaic (PV) arrays. It includes DC array wiring, electrical protection devices, switching and earthing provisions. Load side is not covered. An exception to this is that provisions relating to power conversion equipment only where DC safety issues are involved.
- **IEC 62716: 2013** is the standard for ammonia corrosion testing.
- **IEC 62759-1:2015** is the standard for transportation of PV modules.
- **IEC 62804-1-1:2020** standardizes procedures to test and evaluate for potential-induced degradation-delamination mode in the laminate of silicon PV modules.

- **IECEE TRF 61646E:2018** defines design qualification and type approval of thin-film terrestrial photovoltaic modules
- **IEC 61277, IEC 61345, IEC 61646** are withdrawn standards for PVs.

2.3.2   Manufacturer Technical Specifications

Photovoltaic manufacturers measure and indicate the technical specifications of a PV module on a label which is on the backside of the module. Maximum power, normal operating cell temperature, short circuit current, open circuit voltage are some parameters in these technical specifications. When dimensioning a solar energy system, these parameters need to be correctly understood. Some common parameters on manufacturers' labels, their units and definitions can be found in Table 2.3. These values are measured under AM1.5 conditions. The test conditions are defined as an irradiation of 1000 W/m^2 and a temperature of 25 °C. AM stands for air mass, the thickness of the atmosphere, which in Europe it is approximately 1.5. It should be never forgotten that the solar arrays rarely operate under the standard test conditions and generally the solar irradiation is lower and the cell temperature is higher than in tests.

In addition to these parameters the *fill factor* definition is used as one of the measures that shows the quality of a module. *Fill factor (FF)* is the ratio of maximum power of a module to the product of the open circuit voltage to the short circuit current (Eq. 2.1).

$$FF = \frac{P_{max}}{V_{oc} \cdot I_{sc}} \tag{2.1}$$

Let us remember the photovoltaic cell characteristic graph once more to understand these specifications. P_{max} and (V_{oc} x I_{sc}) areas can be seen in Fig. 2.6.

Table 2.3 Manufacturer technical specifications of a PV module

Specifications		
P_{max}	Wp	Maximum obtainable power, maximum power rating
V_{oc}	V	Measured voltage between the terminals of the module when there is no connection
I_{sc}	A	Measured current when the terminals of the module are short-circuited.
V_{mp}	V	Voltage @maximum power
I_{mp}	A	Current @maximum power
$NOCT$	°C	Normal operating cell temperature
η	%	Module efficiency
		Cell type (mono/poly Si)
c_{Voc}	%/°C	Temperature coefficient of V_{oc}
c_{Isc}	%/°C	Temperature coefficient of I_{sc}

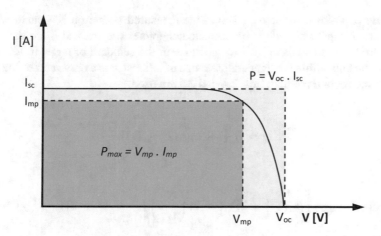

Fig. 2.6 Technical specification parameters of a PV cell

EN 50380 defines the datasheet and nameplate information of photovoltaic modules.

2.3.3   Measuring a Photovoltaic Module

Photovoltaic manufacturers indicate the technical specifications of PV modules on backside labels. However, these specifications are determined at standard test conditions (air mass: AM1.5, 1000 W/m², 25 °C). Photovoltaic cells performance varies with ambient conditions and their efficiency decreases over the years. Thus, their performance need to be determined in actual operating conditions. A data acquisition and logging system is strongly recommended to determine if there are PV modules that generate less energy than they should. Standards consider one – diode model of the PV cell to make corrections about the performance of PV modules (Table 3.1).

Corrections are defined in IEC standards 60891, 60904. For these corrections, there are prerequisites. The global irradiance must be at least 800 W/m², the measuring devices must have an accuracy of ±1% and the device measuring the global irradiation must be within 2 degrees of the testing module. The V_{OC} and I_{SC} measurements must have an accuracy of ±0.2% and must be measured with a four-wire connection. The first of the two common correction methods uses Eqs. 2.2 and 2.3.

$$I_2 = I_1 + I_{SC} \left(\frac{G_2}{G_1} - 1 \right) + \alpha \, (T_2 - T_1) \tag{2.2}$$

$$V_2 = V_1 - R_S \, (I_2 - I_1) - \kappa \, I_2 \, (T_2 - T_1) + \beta \, (T_2 - T_1) \tag{2.3}$$

In these equations, subindex 1 shows the measured points on I-V curve whereas subindex 2 represents points on corrected characteristics. α and β are the voltage and current coefficients of the test specimen in the standard or target irradiance for correction and within the temperature range of interest. κ is a curve correction factor.

For the second method, Eqs. 2.4 and 2.5 are used.

$$I_2 = I_1\ [1 + \alpha_{rel}\ (T_2 - T_1)]\ \frac{G_2}{G_1} \qquad (2.4)$$

$$V_2 = V_1 + V_{OC1} \left[\beta_{rel}\ (T_2 - T_1) + \alpha\ \ln\left(\frac{G_2}{G_1}\right) \right] - R'_s\ (I_2 - I_1) - \kappa'\ I_2\ (T_2 - T_1) \qquad (2.5)$$

In the equations above, V_{OC1} is the voltage at test conditions. α_{rel} and β_{rel} are the relative current and voltage coefficients of the test specimen measured at $1000\ \mathrm{W/m^2}$. R'_s is the internal series resistance of the test specimen. κ' is the temperature coefficient of the internal series resistance R'_s.

2.4 Maximum Power Point Tracking

The current of the PV cell varies with the solar radiation falling on the cell. Since the power is the product of current and voltage; a change in current causes a corresponding change in power output. A constant DC/DC converter has a disadvantage in varying solar irradiation levels since the maximum power point changes.

Loads, voltage level of battery packages, global irradiation (so the output current of the array) are all changing. To utilize the solar array in an efficient way, the maximum power generation point of the PVs needs to be tracked (Figs. 2.7 and 2.8).

In Fig. 2.8, load line, PV output and power curve are represented in the same graph. Consider that the current axis for the power curve is scaled to make the different curves and lines clear. The dashed line represents the current output of the PV module whereas the continuous black line is the power curve. Depending on instantaneous power needs, the load line changes. The intersection point of load line and the source curve is called the operating point. In most cases, the operating point is not a maximum power point (MPP). This means that the PV module is not used efficiently enough and there is a need for a controller to correct this. The device that tracks the maximum power point is called as maximum power point tracker (MPPT). There are different algorithms and topologies to perform maximum power point tracking.

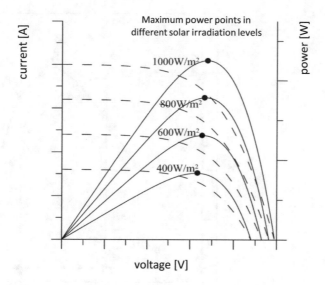

Fig. 2.7   Maximum power point of PVs

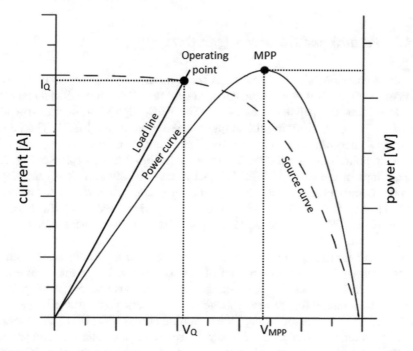

Fig. 2.8   Operating point and MPP

Fig. 2.9 MPP tracking

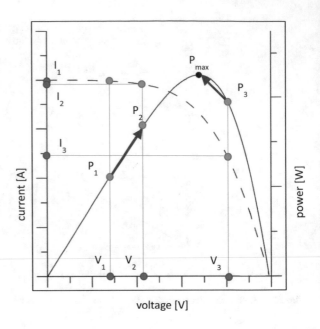

voltage [V]

2.4.1 *Perturb and Observe Algorithm*

The most common MPPT algorithm is the hill climbing algorithm. Perturb and observe (P&O) algorithm, incremental conductance (IC) and incremental resistance (IR) algorithms are popular ones [4, 7]. The P&O Algorithm aims to reach the top of the curve by changing the output voltage value of the MPPT. The perturb and observe algorithm is also known as hill climbing algorithm because finding the maximum point is like a hill climbing from the left or right sides (pls. see Fig. 2.9).

The slope of the curve at MPP is equal to zero. IC algorithm uses the instantaneous conductance, I/V, and the incremental conductance, dI/dV, for detecting MPP [7]. At MPP, $dI/dV = -I/V$. Thus, IC algorithm seeks for the condition $I+V(dI/dV) = 0$. The output of the MPP algorithm is implemented through the PWM voltage output.

A perturbation is applied to the array voltage. The algorithm starts with the current measured voltage and current of the solar array. Using these values, the power is calculated. Then, a new voltage output value is chosen (see Fig. 2.10). The current is sensed for the new voltage value and new power value is calculated once more. In the observation step, the algorithm checks if the new power value is greater than the previous one. If the new power value is greater than the previous one, the baseline voltage value is replaced with the new one. However, if the new voltage value results in a lower power value, then the delta value is subtracted from the initial voltage value and new power value is calculated. The new power value is compared with the previous one and the whole process continues repeatedly.

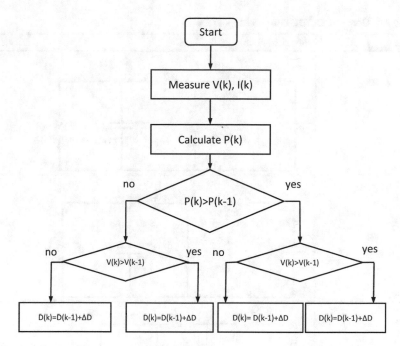

Fig. 2.10 P&O algorithm

This method, however, has a disadvantage of not being able to track fast changing solar irradiation. There are hybrid techniques to overcome this disadvantage. For instance, let us assume that we are at P_1 in Fig. 2.9. During implementation of P_2 in the algorithm, if the solar radiation changed to cause the MPP not to be on the right but on the left of the calculated point, there is a delay to find the MPP. Adaptive control techniques or hybrid methods can overcome this issue.

2.4.2   Converter Topologies

A DC/DC converter is an electronic circuit which increases or decreases voltage or current due to output side needs. There are common topologies that are used in MPPTs [6, 8]. These are, buck, boost, buck-boost and cuk types of DC/DC converters. In Table 2.4, the PV array side is denoted as input voltage (V_I) and load is connected to output voltage (V_O). The switch in this table simulates the switching in the DC-DC conversion, so MPPT algorithms need to be analyzed in two states: When the switch is at the "on" state and when it is "off". In Fig. 2.11, the states of "buck type" topology in the two states (during operation) can be seen.

A buck type DC/DC converter is a voltage step down converter. Since, the power is the product of voltage and current, the current is stepped up in buck converters.

Table 2.4 Common DC/DC converter topologies

Type	Electrical circuit
Buck	
Boost	
Buck – boost	
Ćuk	

When the switch in Fig. 2.11 is on-state the current increases, so the inductor produces an opposing voltage in response to the current change. This reduces the voltage across the load. When the state of the switch is off, the voltage source is disconnected from the circuit and the current decreases. The changing current produces a change in voltage through the inductor. During this time interval, the inductor discharges its energy to the circuit. In buck type MPPTs solar array has higher voltage level than load side whereas in boost type it is lower (Fig. 2.12).

A boost type DC/DC converter is a voltage step up converter which means also that the output current is lower than the input current. When the switch in Fig. 2.12 on-state, current flows through the inductor and the inductor stores some energy by generating a magnetic field. When the switch is at the off-state, the increase

Fig. 2.11 Buck type MPPT
topology during operation

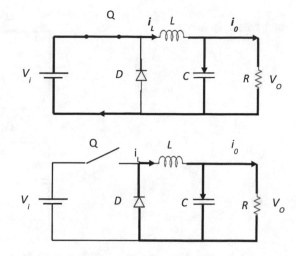

Fig. 2.12 Boost type MPPT
topology during operation

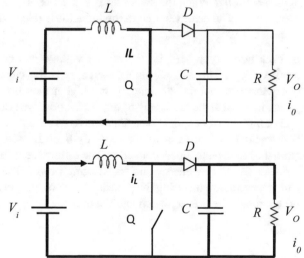

in impedance reduces the current. To maintain the current through the load, the magnetic field is used. This causes the polarity to be reversed, which causes the two sources to be in series and a higher voltage to charge the capacitor through diode D. In boost type MPPTs, the load side or the output voltage is higher than the solar array side. On the other hand, in buck-boost type (pls. see Fig. 2.13), the output voltage level can be lower or higher than input side.

A buck – boost type MPPT is equivalent to a flyback converter using a single inductor and the topology is a combination of buck and boost types.

Switch on state: When the switch in Fig. 2.13 is on, all the current flows through the switch and the inductor and returns back to the source.

Fig. 2.13 Buck – boost type
MPPT topology during
operation

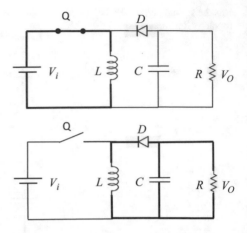

Switch off state: When the switch in Fig. 2.13 is off, the polarity of the inductor is
reversed. The energy stored in the inductor is released and dissipated on the load
resistance.

In buck-boost type MPPTs, the output voltage can be greater or lower than the
input voltage. This has advantages such as being capable of covering entire I-V
characteristics, but the efficiency is not higher than buck or boost types. When the
power increases, the efficiency of the buck-boost type decreases dramatically.

For high power applications, Cuk converters are more efficient. A Cuk converter
is a negative-output capacitive energy fly-back DC–DC converter. It is developed
from the basic buck–boost converter. The difference is that it uses a capacitor not an
inductor for storing energy and transferring power. The main disadvantages of the
Cuk converter are inverting the polarity of the output, in addition to the high current
flow through the power switch and the diode.

2.4.3   Common MPPT Devices

The selection of structure and topology of MPPTs are dependent to technical
parameters like power of sources, types of sources, power of loads, types of loads,
nominal voltage levels and the additional performance that the MPPT can provide.
In applications, the system is shaped generally by the budget. If it is a simple system
with relatively smaller loads which are activated during short time intervals (like
sensor implemented lighting or alarm systems or a stand-alone system on a hut
roof which is not used more than one or two days in a week), PWM solar charge
controllers can be used for charging a battery. PWM solar charge controllers are
cost-effective solutions for off-grid systems. In Fig. 2.14, a solar charge controller
(a), an MPPT solar charger (b) and a hybrid charge controller (c) can be seen. Since
the nominal voltages of lead acid batteries are multiples of 2 V, it is common to

Fig. 2.14 Examples of a PWM solar charge controller, an MPPT solar charger and a hybrid MPPT controller; (**a**) PWM solar charge controller, (**b**) MPPT – Victron BlueSolar 150/70 (**c**) Wind solar hybrid MPPT – Marsrock 1000 W solar, 1000 W wind, 12 V/24 V/48 V

find lead acid batteries in 12 V packages, generally the output voltage levels of residential MPPTs can be adjusted to 12 V, 24 V or 48 V. Most of the MPPT solar chargers and even simple PWM solar charge controllers can sense the number of serially connected batteries when the battery pack is connected to power the MPPT.

Deep cycle lead acid batteries are very common in residential systems. Flooded, absorbed glass mat and gel types are deep cycle lead acid types. These type of batteries are heavier than most other types of batteries. Nevertheless, they are cost effective, robust and have a longer life. For these reasons, they are used in marine applications or in caravans, too.

On the other hand, residential MPPT devices can combine wind energy or other energy sources with solar energy. An on-grid hybrid MPPT unit and its connections can be seen in Fig. 2.15. Most hybrid on-grid MPPTs are not connected to batteries. If a battery is connected, an additional dump load is connected to hybrid charge controllers to protect batteries for over charging.

2.5 Inverters

Photovoltaic arrays produce DC electricity, while the distribution and transmission grids mostly operate with AC electricity at 50 or 60 Hz, as do many electrical devices in residential, commercial, and industrial premises. To interface between the PV array and an AC load, or between the PV array and the electricity grid, an inverter is necessary to convert the DC electricity that is produced by the PV array to suitable AC electricity. The alternating current sine waves are created by the inverter's solid state switching devices, such as insulated-gate bipolar transistors (IGBTs), see Fig. 2.16. In some inverters, transformers are used to increase the

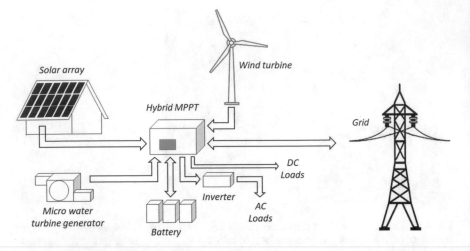

Fig. 2.15 Connections in hybrid MPPTs

Fig. 2.16 The picture shows a Mitsubishi IGBT module rated at 3300 V and 1200 A. (Source: https://commons. wikimedia.org/wiki/File: IGBT_3300V_1200A_ Mitsubishi.jpg. Licensed under CC BY-SA 3.0. Author: ArséniureDeGallium)

AC output voltage of the inverter so that it can be connected to an AC system. Many inverters used in photovoltaic applications incorporate maximum power point tracking (MPPT) functionality.

2.5.1 Principles of Operation

2.5.1.1 Full Bridge and Half-Bridge Inverters

Consider the full-bridge inverter illustrated in Fig. 2.17a, which has four switches labelled P and N. These switches open and close periodically with a period T. When the P switches are closed (ON), the N switches are open (OFF), and vice-versa. The

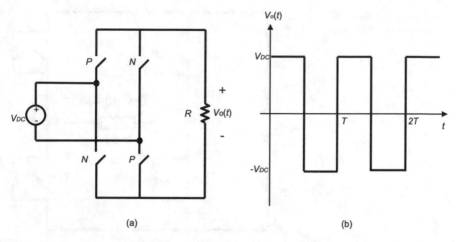

(a)   (b)

Fig. 2.17 (**a**) Circuit diagram for the full-bridge inverter. (**b**) Resulting output waveform from the full-bridge inverter when the P switches open and close with a period T, with a phase shift of $T/2$ from the N switches, which operate at the same frequency as the P switches

Fig. 2.18 Circuit diagram for the half-bridge inverter. The P switch opens and closes with a period T, with a phase shift of $T/2$ from the N switch, which operates at the same frequency as the P switch

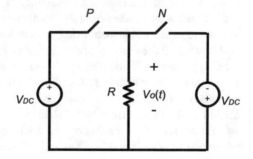

ratio of the ON time of the switch to the period T is defined as the duty ratio or duty-cycle (D). Suppose that both P and N switches operate with $D = 0.5$. The resulting AC waveform (a square wave) is illustrated in Fig. 2.17b.

A half-bridge inverter is designed using two equal DC voltage sources, as well as two switches labelled P and N, as illustrated in Fig. 2.18.

Clearly, the output waveforms of the full-bridge and half-bridge inverters contain harmonics. Inverters use different mechanisms for eliminating or reducing these harmonics. For example, a passive low-pass filter, such as an LC circuit, can be designed to eliminate most of the harmonics that are part of the square wave waveform, in such a way that the output has very little harmonic content.

2.5.1.2 Full-Bridge PWM Switched Unipolar Inverter

The method of pulse-with modulation (PWM), which is used in many inverter designs, introduces several zero voltage regions at a high frequency within a single

Fig. 2.19 Full-bridge PWM
switched unipolar inverter

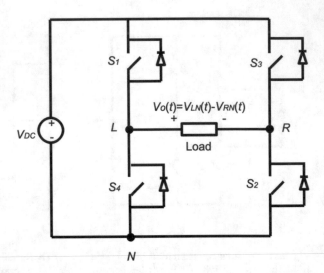

period of the fundamental AC frequency. This results in an output waveform that only contains the fundamental AC component, as well as high frequency harmonics which are easier to filter out using passive low-pass filtering.

Figure 2.19 illustrates a converter in full-bridge PWM switched unipolar inverter topology, where anti-parallel diodes are added to allow for bidirectional conduction of the switches, which is required with inductive loads. In this topology, the pair of switches on the left-hand-side (S1 and S4) operate at a high frequency that is known as the carrier frequency, which is significantly higher than the fundamental AC frequency, while the other two switches on the right-hand side (S2 and S3) operate at the fundamental frequency. Note that during each half-cycle of the low-frequency switches, the high-frequency switches operate with variable ON-times. This variable duty cycle can be obtained by comparing a triangular wave V_t at the carrier frequency with a sine wave at the fundamental frequency, which is also called the control signal V_c, or it can be generated by a microcontroller. Switch S_1 closes when $V_c > V_t$, S_2 closes when $V_c > 0$, S_3 closes when $V_c < 0$, and S_4 closes when $V_c < V_t$. This switching pattern results in an output waveform V_o with a constant frequency and a modulated pulse width that is illustrated in Fig. 2.20.

The amplitude modulation factor, m_a, of the PWM is defined as the ratio of the amplitude of the control signal V_{cM} to the amplitude of the triangular wave V_{tM}, noting that $m_a \leq 1$.

$$m_a = V_{cM}/V_{tM} \tag{2.6}$$

The frequency modulation factor, m_f, is defined as the ratio of the carrier frequency f_c to the ratio of the fundamental frequency f_1, noting that $m_f \gg 1$.

$$m_f = f_c/f_1 \tag{2.7}$$

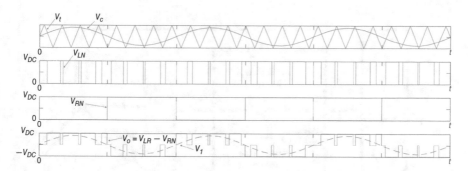

Fig. 2.20 Waveforms in a full-bridge PWM switched unipolar inverter, including the triangular waveform V_t at the carrier frequency, the control signal V_c at the fundamental frequency, the voltages at the left node of the bridge V_{LN}, the voltage at the right node of the bridge V_{RN}, the output voltage V_o, and the first harmonic of the output voltage V_1

It is not difficult to show that the amplitude of the first harmonic of the output signal, V_1, is given by [3]

$$V_{1M} = m_a V_{DC} \tag{2.8}$$

After the first harmonic at the fundamental frequency, the next harmonic of the output signal that occurs has a frequency of around $2m_f\, f_1$, and the next harmonic after that has a frequency of about $4m_f\, f_1$.

2.5.1.3 Three-Phase Inverters

Three-phase inverters can be made using single-phase inverters. Three single phase inverters can be connected to a single input DC voltage. Their low frequency switching controls are shifted by 120 degrees. Figure 2.21 illustrates the configuration of a three-phase inverter made up of three single-phase inverters connected in Y.

It is also possible to construct an inverter using a three-phase bridge topology [3], which consists of three pairs of bilateral switches, as illustrated in Fig. 2.22. The pairs of switches in each branch (S1, S4), (S3, S6), and (S2, S5) are operated in such a way that when one switch is closed, the other member of the same pair is open.

PWM is also commonly used in three-phase bridges. In this case, the frequency of the fundamental harmonic of each phase is equal to the frequency of the sinusoidal control signal. The amplitude of the fundamental harmonic can be controlled by setting the amplitude modulation ratio, m_a. Higher harmonics are present at the frequency of the carrier signal and its integer multipliers, which makes it easy for them to be filtered. Each pair of the switches requires one control signal, and the three control signals are phase shifted by 120°.

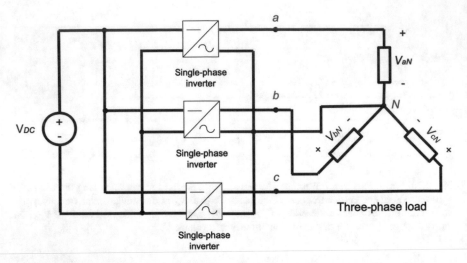

Fig. 2.21   Illustration of the three-phase inverter made up of three single-phase inverters connected in Y

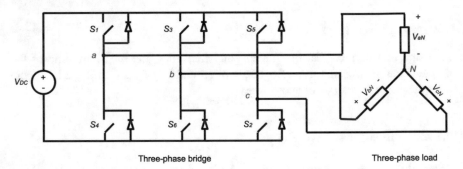

Fig. 2.22   Three-phase bridge inverter involving three pairs of bilateral switches with a three-phase load in Y connection

2.5.2   Grid-Tie and Off-Grid Inverters

A grid-tie inverter converts DC electricity into AC electricity suitable for injecting into an electrical power grid. To inject electrical power efficiently and safely into the grid, grid-tie inverters must accurately match the voltage and phase of the grid sine wave AC waveform. A grid-tie inverter senses the current AC grid waveform, and outputs a voltage to corresponds with the grid's voltage both in magnitude and phase. Typically, grid-tie inverters operate with a unit power factor, such that the inverter's voltage and current are in phase. However, supplying reactive power to the grid might be necessary to keep the voltage in the local grid inside allowed limitations and many grid-tie inverters have this capability. Grid-tie inverters are also designed to quickly disconnect from the grid if the utility grid goes down.

An off-grid inverter is, as the name implies, a solar inverter that is not connected the AC electricity grid, meaning that it works alone and cannot be coupled in any way with an existing AC grid.

2.5.3 Commercial Solar Inverters

Modern day commercial solar inverters are sophisticated devices that not only convert the DC electricity from solar arrays into AC electricity but can also provide additional functionality including maximum power point tracking, filtering, monitoring, communications, and user interfacing. They can range in terms of the power they can handle from a few hundred watts in the case of micro-inverters, to several megawatts in the case of large central inverters used in utility scale solar farms.

Figure 2.23 illustrates the components of a typical solar inverter. The DC/DC converter raises or lowers the incoming DC voltage from the PV array, implementing MPPT to extract maximum power from the PV array. The next stage is the actual power inverter, which is driven by a microcontroller with multiple PWM outputs to control the solid-state power switches. The inverter can be isolated or non-isolated. Galvanic isolation can be achieved by the means of a transformer, which has a negative impact on a grid-connected PV system's efficiency. The efficiency of PV inverter systems can be improved by using transformerless topologies [5]. The microcontroller has an onboard analogue to digital converter to monitor critical parameters of the system, such as DC and AC currents and voltages. The electromagnetic interference (EMI) filter on the DC side helps reduce conducted electromagnetic interference towards the PV array and decouples the PV array and the inverter. The EMI filter on the AC side reduces the harmonic content injected into the AC grid. Many inverters have a user interface with a display and buttons to enable interaction with the user. Many inverters have data communications capabilities usually through ethernet or WiFi connections that are intended to allow remote monitoring of the system via a cloud service provided by the manufacturer. Figure 2.24 shows an image of the internal components of a commercial solar inverter.

2.6 Standalone PV Systems

Photovoltaic systems are often classified into two groups as on-grid systems and off-grid systems. Off-grid systems are the ones that are not connected to the AC electricity network. Standalone systems are small systems which are off-grid. Forms of output of these systems depends on power input needs of loads.

Depending on the distances between sources and loads, budget and total efficiency of the system, the output voltage of the stand alone PV systems can be

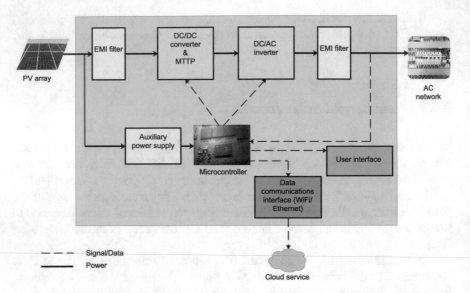

Fig. 2.23 This diagram illustrates the typical components and additional support functions of a solar inverter

either AC or DC form. If the voltage output of the system is in DC form, the cable diameters are wider than AC form selection. This causes an increase in cable cost. In addition to this, comparing to AC loads with the same power, DC loads are more expensive. On the other hand, if the output voltage form is selected as AC, an inverter cost is added to the system.

In Fig. 2.25, an illustration of a stand-alone PV system can be seen. Nominal system voltages for stand-alone systems are supposed to have 12 V, 24 V or 48 V since these are the multiples of lead acid battery nominal voltages. Another advantage is that it is easier to find some loads, circuit breakers or components using these voltages since the vehicle nominal voltages are also 12 V or 24 V.

Standalone PV systems are suitable for the locations where the grid is not easier to reach or very expensive to connect. Some applications are: bus stops, telecom base stations, rural houses, rural research stations, irrigation, public lighting and other public loads. In order to reduce the total investment cost of the system, it is a rational choice to use if there is an energy storage facility other than batteries. For example, in irrigation systems, it is common and cheaper to store the energy as the potential energy stored in an elevated water tank. Another passive form of energy storage is in the form of heat, such as hot water tank.

Fig. 2.24 This picture shows the actual internal components of a SMA SUNNYBOY 3000 solar inverter, which is a 3-kW inverter used for residential applications. (Source: https://www.flickr.com/photos/rneches/2541181118/. Licensed under CC BY 2.0. Author: Russel Neches)

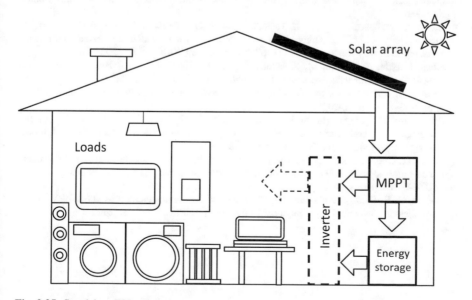

Fig. 2.25 Standalone PV system

2.7 Grid Connected PV Systems

A grid connected PV system is one where the photovoltaic array is connected to the utility grid through a grid-tie inverter, allowing the system to operate in parallel and exchange energy with the electric utility grid. Grid-connected systems are found in areas that have a readily available AC electricity network. Some large photovoltaic plants are only intended to inject power into the electricity grid, while other PV systems serve a local load (such as a building) and may export power into the grid if there is an excess of generated solar energy. In this way, the grid acts as a kind of storage medium and when power is needed locally it can be imported from the grid. It is intended that the PV system does not have to supply enough electricity to cover the local power demand as in an off-grid system. The local demand can be supplied by the PV system, the electricity grid or a combination of the two, such that the system can be as small or large as may be desired. The major components of a grid connected PV system include the PV array, inverter, the necessary cables, protection devices, metering and monitoring systems, switches and transformers.

References

1. Becquerel, E. (1839). Mémoire sur les effets électriques produits sous l'influence des rayons solaires, C. R. Seances Acad. Sci. D, 9, pp. 561–567.
2. Guarnieri, M. (2015). More Light on Information [Historical], IEEE Industrial Electronics Magazine, 9(4), pp. 58–61, Dec. 2015, https://doi.org/10.1109/MIE.2015.2485182.
3. Dokić, B.L. and·Blanuša, B. (2015). Power Electronic Converters and Regulators. Third Edition. Springer.
4. Jately, V., Azzopardi, B., Joshi, J., Venkateswaran V. B., Sharma, A., Arora, S. (2021). Experimental Analysis of hill-climbing MPPT algorithms under low irradiance levels, Renewable and Sustainable Energy Reviews, 150, 111467, https://doi.org/10.1016/j.rser.2021.111467.
5. Khan, M.N.H, Forouzesh, M, Siwakoti, Y.P, Li, L., Kerekes, T. and Blaabjerg, F. (2020). Transformerless Inverter Topologies for Single-Phase Photovoltaic Systems: A Comparative Review, in *IEEE Journal of Emerging and Selected Topics in Power Electronics*, vol. 8, no. 1, pp. 805–835, March 2020,
6. Ravindranath Tagore, Y., Rajani, K. & Anuradha, K. (2022) Dynamic analysis of solar powered two-stage dc–dc converter with MPPT and voltage regulation. Int. J. Dynam. Control, https://doi-org.libproxy.viko.lt/10.1007/s40435-022-00930-8.
7. Shang, L., Guo, H. & Zhu, W (2020). An improved MPPT control strategy based on incremental conductance algorithm. Prot Control Mod Power Systems, 5(14). https://doi.org/10.1186/s41601-020-00161-z
8. Walker, G. R. (2000), Evaluating MPPT Converter Topologies using a MATLAB PV Model, Australiasian Unversity Power Engineer Conference, AUPEC, Brisbane, 2000.

Chapter 3
Battery Technologies

3.1 Introduction to Batteries

Energy storage is a method of storing energy produced at one time to be used at some point in the future. Energy storage technologies are diverse, and as are their principles of operation and effectiveness. The main types of energy storage are:

- Mechanical: compressed air energy storage, flywheels, pumped storage hydropower
- Electrical: capacitors and superconducting magnetic
- Thermal: sensible heat, latent heat, cryogenic, Carnot battery
- Chemical: biofuels, hydrogen, methane, power-2-gas/liquid
- Electrochemical (batteries): rechargeable battery, flow battery, supercapacitor

Five main energy storage types are radically different, both in terms of the quantity of energy stored and storage times [3]. The selection of the energy storage method and its associated technology is dependent on the application. The main types energy storage technologies are compared in terms of their storage capacity and discharging times in Fig. 3.1.

Mechanical energy storage solutions concern mainly the Pumped Storage Hydropower (PSH) technology that is used to store large quantities of energy, and the Compressed Air Energy Storage (CAES) technology to a lesser extent. Flywheel technology can be used to store smaller amount of energy but benefits from a very fast discharging time that have a particular interest for transport applications.

The primary electrical energy storage method is the use of capacitors to store very small amounts of electricity, mostly for electronic devices. The new superconducting magnetic energy storage (SMES) technology, also used for short term storage, is still under development.

Thermal energy storage is mostly used for applications where heat is stored from around an hour to a few months, such as the domestic hot water or concentrated solar power plants storage solutions.

© The Author(s), under exclusive license to Springer Nature Switzerland AG 2023
A. Rachid et al., *Solar Energy Engineering and Applications*, Power Systems,
https://doi.org/10.1007/978-3-031-20830-0_3

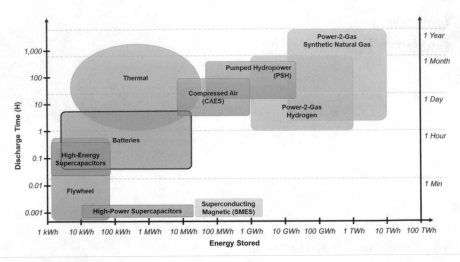

Fig. 3.1 Comparison of storage capacity and discharge times for different energy storage technologies

Chemical form includes technologies like hydrogen or biofuels, where large amounts of energy are stored for a long period of time.

Electrochemical solutions, known as batteries, can store part or excess electricity and feed it back in times of high demand and as such are naturally paired with intermittent renewable energy systems, namely solar and wind.

Almost all (96%) of the grid connected operational stored capacity in the world is under based on pumped-storage hydropower technology (170 GW). The rest of the operational stored capacity is distributed between the thermal, other mechanical and electrochemical storage that represents around 2 GW (Fig. 3.2). Several different battery technologies are employed in grid-connected electrochemical energy storage, developed within the last few decades. The most dominating technology for all types of applications, is the lithium-ion battery with almost 80% of the global capacity (Fig. 3.3).

In 2020, this trend continued with Lithium-ion batteries representing almost 90% of the electrochemical storage, with installed capacity rising to over 9 GW, which corresponds to an increase by a factor of 6! With the total grid connected battery storage capacity of to more than 10 GW, the battery market in the world is immensely dynamic and ever growing.

Batteries are easy to install and during their operation have minimal carbon emissions. Different battery types are discussed below.

Fig. 3.2 Grid connected total
operational capacity
worldwide in 2018 [8]

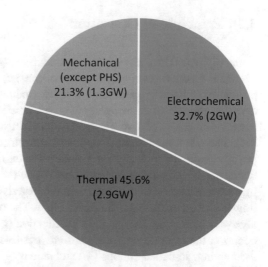

Fig. 3.3 Grid connected battery storage

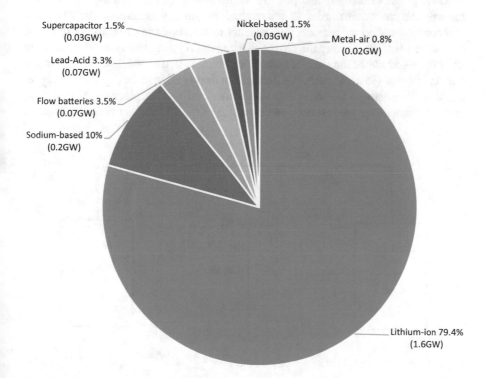

3.2 Principle of Operation

Generally speaking, the electrochemical energy storage is mainly based on a chemical reaction involving electron exchange, or in other words, electricity. The principles of battery operation are redox (reduction-oxidation) chemical reactions, allowing the use of exchanged electrons, or electricity, for another application. A battery comprises of two electrodes, an electrolyte and an electrical circuit (Fig. 3.4). The electrical circuit links the two electrodes, which are always polarized as positive (cathode) and negative (anode). The two electrodes sandwich an electrolyte, which can be a liquid, a paste or a gel. Once the battery is in operation, the current flows from the positive electrode to the negative one. By convention, the electron flow is always opposite to the current flow. The oxidation (electron release) reaction always occurs at the anode, and the reduction (gaining an electron) reaction always occurs at the cathode. In the electrolyte, positively and negatively charged particles are moving, so-called cations (+) and anions (−), respectively.

During the discharge, the positive components, called cations, are moving towards the cathode in order to get reduced by gaining electrons. The negatives components, called anions, are then moving to the anode in order to get oxidized by losing electrons. During the charge, the current is going from the positive to the negative electrode of the current generator (which is in the middle). Then from the battery point of view, the current is going from the negative to the positive electrode, and then the electrons are going in the opposite direction. The place where oxidation and reduction occur are inverted, and the name of the electrodes are also changing.

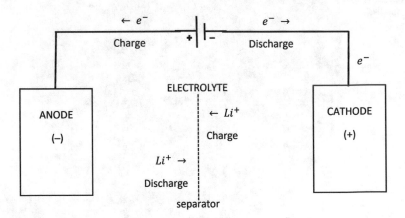

Fig. 3.4 Working principle of a Lithium ion battery

3.3 Applications of Battery Storage in PV Systems

Battery storage is an essential tool for managing a mismatch in demand and production. If we take a basic example of a typical house with solar panels, majority of solar electricity is produced during the day when the sun is shining and the demand (self-consumption) is low. Thus, the excess of electricity produced is exported to the grid. Conversely, peaks in consumption tend to be early in the morning or late at night when production is low or impossible. By comparing the consumption curve of the house and the production curve of the solar panels we can easily identify the mismatch (Fig. 3.5).

In order to avoid exporting the energy and maximise self-consumption of the solar energy, a storage system is required to store a part of the solar electricity production during the day to be used later when needed, especially during the consumption peak in the evening and in the morning. Therefore, the principal aim of battery storage for solar applications is to manage the source/demand shift. However, managing the mismatch between the production and the consumption is more complex as in reality, on the consumption side, each consumer may have different energy needs and consumption curve, which often changes across the year. On the production side, the intermittent nature of the solar resource presents a challenge, whether it is predictable like day/night cycles or seasons or unpredictable like clouds.

3.4 Battery Characteristics

When selecting a battery for certain application, a number of attributes must be considered. These are known as battery characteristics – technical quantitative parameter describing battery performance. For solar energy storage, battery efficiency and capacity, charging and discharging, useful life and operating temperature, as well as battery size and weight are essential.

Size and weight of the battery are important considerations. Energy density is the amount of energy stored per volume of the battery, expressed in Wh/L. Specific energy is the energy stored per mass of the battery, expressed in Wh/kg. Power density is the amount of power stored per mass of battery, expressed in W/kg.

Efficiency is a very important parameter to consider when selecting a battery. It shows percentage of energy taken from the battery during discharge, compared to the energy directed to the batter during charging.

Battery capacity is the amount of energy stored through electrochemical reactions in the battery, measured in ampere-hours (Ah). Battery capacity is sometimes measured in Watt-hours (Wh) or kilowatt-hours (kWh), estimated by multiplying ampere-hours by the operational voltage of the battery in volts (V). For applications where battery voltage changes during charging and discharging, this can give an imprecise value. Battery nominal rated capacity is significantly affected

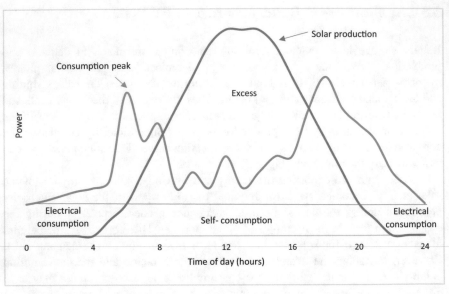

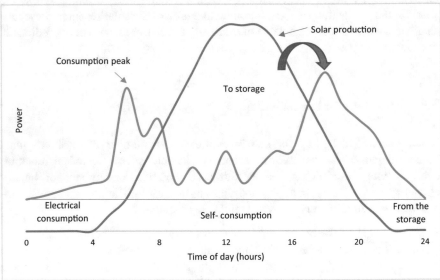

Fig. 3.5 Typical consumption and production curves without storage (above) and with storage (below)

by temperatures and the rate of charging, and for detailed evaluation it is always best to consult the manufacturer specification.

Rate of charging and discharging affect the battery performance. Charging/ discharging rates are nominally measured in amperes. However, it is more to specify the rate as the amount of time needed to fully discharge the battery, even though in practice the battery would never be fully discharged as this can lead to reduced life and poor battery performance. For example, fast discharge means that the current value is high, and the amount of energy taken from the batter is lower; hence, the capacity is reduced. Manufacturer specifications would normally give the battery capacity a function of time needed to fully discharge the battery. Advised charging and discharging rates are also stated.

Operational temperature also affects battery capacity. Higher temperatures typically lead to higher capacity, but this decreases the useful life of the battery.

Depth of discharge (DOD) is the amount of energy taken from the battery, expressed as a percentage of the total capacity. Complete discharge is not advisable for majority of battery types. If battery has been discharged below the nominal DOD, the capacity and the nominal life of the battery will be reduced. However, even when advised DOD is used, over time the battery capacity diminishes. For PV installations, daily depth of discharge (max amount of energy that can be taken form a battery in 24 h) is an important consideration.

Battery State of Charge (BSOC) is the percentage of energy stored in the battery compared to the total battery capacity. It is often expressed as voltage of the battery compared to the voltage of a fully charged battery. However, battery voltage depends on the temperature.

Battery life is normally expressed in numbers of charging/discharging cycles during which the battery keeps a certain capacity. Gradual degradation of the materials in the battery during electrochemical reactions reduces the length of time the battery can be used. Battery life can be expressed in years, which is useful for systems which are not frequently charged and discharged. For PV applications, battery life is measured in the number of charge/discharge cycles, and depends on temperature and DOD.

Safety regulations must be checked for the specific type of battery used. Most batteries contain harmful (toxic and/or corrosive) chemicals, and electrical safety protocols must always be adhered to. Batteries must be disposed of safely and recycled whenever possible. Sealed batteries do not require maintenance. However, batteries where reaction product change the volume of the battery require regular maintenance.

3.5 Battery Types

Leading battery technologies used to store electricity in solar applications include lead-acid batteries, nickel-based batteries, lithium-ion batteries and flow batteries. These technologies are compared and contrasted based on their underlying chemistry (materials and reactions), technical aspects (performance, operating temperature, lifetime, and cost), environmental impact and their applications.

3.5.1   Lead-Acid Batteries

General Characteristics

The lead-acid battery is the first type of rechargeable battery created in 1859 by the French physicist Gaston Planté. It is composed of a lead and lead oxide electrodes and aqueous sulfuric acid as an electrolyte. Compared to newer types of rechargeable batteries, lead-acid batteries have one of the lowest energy densities. Despite this, their ability to deliver high currents means that the cells have a relatively high power-to-weight ratio. These characteristics, along with their low cost, make them attractive for use in motor vehicles to provide the high current required by starter motors. Because they are inexpensive compared to newer technologies, lead-acid batteries are still today widely used even when other designs could provide higher energy densities.

Large-format lead-acid designs are widely used for storage in backup power supplies in cell towers, high-availability environments such as hospitals and stand-alone power systems. For these roles, modified versions of the standard cell can be used to improve storage times and reduce maintenance requirements. Employing lead-acid batteries for utility applications, such as peak shaving, is not feasible. Due to nature of this application deep discharges occur, which as previously highlighted shorten a lead-acid battery's life. Hence, the use of lead-acid batteries for utility applications is not extensive and therefore they only have limited use in solar power storage [5].

Nevertheless, the weak point of this technology is the formation and dissolution of lead sulfate on the pure lead negative electrode over the charge and discharge cycles, which is clearly decreasing its performance with age and limiting its lifetime. New lead-acid systems have carbon added to this electrode to overcome this problem. Lead-Carbon batteries are further discussed below.

The biggest issue with lead-acid batteries occurs at the environmental level. Lead is highly toxic and recycling it can lead to pollution and contamination, resulting in many lasting health problems. However, lead recycling is one of the easiest to achieve, where up to 100% of lead-acid batteries can be recycled and reused for a second life, which in this sense limits its relative impact to the environment. As an example, 99% of lead-acid batteries in US have been recycled between 2014 and 2018. But this ratio is in reality fluctuating from 60% to 95% since 1982.

Chemistry of Battery

Anode $(-)$: $Pb + SO_4^{2-} \leftrightarrow PbSO_4 + 2e$
Cathode $(+)$: $PbO_2 + 4H^+ + SO_4^{2-} + 2e^- \leftrightarrow PbSO_4 + 2H_2O$

Due to being inexpensive in comparison to alternative batteries, whilst having a high peak-power, lead-acid batteries are one of the most common forms of battery storage, in terms of kWh. The open circuit voltage of a lead-acid battery is 2.1 V. Like all batteries, the service life is significantly affected by the battery's usage. Deep discharges and/or over-charging both reduce a battery cycle life. In stand-by power applications, service life can be more than 10 years [12] and batteries built for cycling can have a cycle-life of 800 cycles. Lead-acid batteries have an operational temperature of -10 to $40\ ^\circ C$. At low temperatures of this range they have poor performance and therefore may need a temperature management system. At least 70% of all lead produced in the western world is used in batteries, 75% of which are car batteries, 15% are stationary batteries and the remaining 10% are traction batteries.

Material requirements for the operation of lead-acid batteries on a daily cycling for 20 years (7300 cycles) delivering 150 kWh/day are shown in Table 3.1. Over 20 years, 200 cells will be replaced four times and approximately 25,000 l of water will be added intermittently.

Environmental Impact

At the end of a battery's life the lead in a lead-acid battery can be recycled and reused in an alternative battery. Increasing the use of secondary lead from 50% to 99% in a lead-acid battery, reduces the electricity and primary energy during material production by 43% and 8%, respectively. When a high-quality battery is required up to 60% secondary lead may be used, compared to low performance applications where up to 100% secondary lead can be used.

Lead-Carbon (PbC) Batteries

Lead-Carbon batteries are essentially an improvement built upon the common lead-acid batteries. The addition of Carbon has made this battery an innovative solution to conventional or arising competitors, as lead-acid batteries still lead the market as mentioned previously. A common problem which lead-acid batteries face, is the accumulation of sulphates. These will make the battery life short due to the negative plate being pure lead being dissolved when the battery is charged in

Table 3.1 Characteristics of Lead-acid battery

Specific energy	35–40 Wh/kg
Energy density	80–90 Wh/L
Specific power	140–180 W/kg
Efficiency (charge/discharge)	50%/95%
Self-discharge rate:	3–20%/month
Operating temperature (C)	-10 to $40\ ^\circ C$ (max 45 $^\circ C$, min $-35\ ^\circ C$)
Useful life	500–1000 cycles $\approx$ 4–5 years
Cost	7–18 Wh/US $
Applications	Automotive and traction application Standby/Back-up/Emergency power for electrical installations Lighting High current drain applications Sealed battery types available for use in portable equipment Grid scale energy storage

cycles. In a lead-carbon system, a Carbon electrode is added to the negative plate, turning the battery into a supercapacitor and improving greatly the charge and discharge performance. The Carbon merely facilitates the electric conductivity and the charging and discharging capacities of the battery All materials found in this battery are achievable and cost-effective.

Main advantages of PbC battery are its high energy density, and good max temperature range. It offers rapid charging, whilst the rate of self-discharge is low. PbC is safer than lead-acid batteries, entirely sustainable and fully recyclable. There is no shortage of elements and the cost of materials is low. However, PbC batteries are very heavy and large. Recommended temperature range is narrow and depth of discharge is low. These batteries cannot fully discharge and there is still a risk of being sulphated.

PbC batteries can last for 20+ years, ideally in a partial state of charge (PSOC) cycle applications. The battery is estimated to withstand 3000 cycles at 80% DOD at high temperatures of about 40 °C, so theoretically the life could be much higher if the temperature was kept at approximately 25 °C [16]. Recommended operation temperatures range from 15 to 25 and maximum temperatures from −20 to 50. Usual efficiency is estimated to remain at 80%, average specific energy ranges from 35–50 Wh/kg and energy density is approximately 100 Wh/L [9, 14].

Lead-carbon are fully recyclable at a rate of up to 98%. Lead has to be extracted and manufactured with care as there is a risk of lead poisoning. Even when recycling the process has to be meticulous as there have been several cases of fumes filtrating into crops and water supplies. Similarly to lead-acid batteries, lead-carbon batteries have a very versatile range of applications, many of which include telecommunications systems, electronic bikes and electric vehicles, solar and wind hybrid storage systems, smart power grids and micro grid systems.

3.5.2   Nickel-Based Batteries

The nickel-based batteries used for solar applications mainly concerns the nickel-iron battery and the nickel-cadmium battery.

Nickel-Iron (NiFe) Batteries

General Characteristics

NiFe aqueous batteries were one of the many designed by Thomas Edison in the early twentieth century. They have been used commonly as rechargeable high-rate batteries, and are currently leading the market in systems that use iron electrodes as anodes. They are arising in popularity due to their low toxicity as opposed to lead-acid cells and the price of the raw materials amongst other things. Iron and Nickel are common metals which are easily extracted, making such batteries very promising as there is no risk of element shortage [11].

Table 3.2 Characteristics of NiFe battery

Specific energy	19–25 Wh/kg
Energy density	30 Wh/L
Specific power	100 W/kg
Efficiency (charge/discharge)	<65%
Self-discharge rate:	20–30%/month
Operating temperature (C)	-20 to 40 °C (max 60 °C, min -30 °C)
Useful life	3000–11,000 cycles $\approx$ 20–85 years
Cost	1.5–6.6 Wh/US $
Applications	Previously: Material handling vehicles, underground mining vehicles, miners' lamps, railway cars and signal systems, emergency lighting Nowadays: Wind and solar power systems where weight is not important, off-grid applications, daily charging applications, for Hydrogen production, for fuels cell cars and storage

The nickel-iron battery is composed of a positive nickel (III) oxide-hydroxide plates and a negative iron plate, with a potassium hydroxide electrolyte. This battery was initially designed to be almost indestructible. Its structure is made to prevent any electrical damage due to, for example, overcharging, short-circuiting and repeated deep discharges. It can have a very long life even under such conditions. It is often used in backup situations where it can be charged continuously and can last for over 20 years. The ability of these batteries to survive frequent cycling is due to the low solubility of the reactants in the electrolyte. The formation of metallic iron during charging is slow due to the low solubility of ferrous hydroxide. While the slow formation of iron crystals preserves the electrodes, it also limits high rate performance: these cells charge slowly, and can only discharge slowly. Nickel-iron cells should not be charged from a constant voltage supply because they can be damaged by thermal runaway; the internal cell voltage drops at the start of gassing, which increases the temperature, which increases the current draw and thus further increases gassing and temperature.

Chemistry of Battery

Anode ($-$): $2NiO(OH) + 2H_2O + 2e^- \leftrightarrow 2Ni(OH)_2 + 2OH^-$
Cathode ($+$): $Fe + 2OH^- \leftrightarrow Fe(OH)_2 + 2e^-$

Many rail vehicles use NiFe batteries. But due to its low specific energy, low charge retention and high manufacturing cost, other types of rechargeable batteries have replaced the nickel-iron battery in most applications. Nevertheless, the technology has regained popularity for off-grid applications where daily charging makes it an appropriate technology. Nickel-iron batteries are being studied for use as combined batteries and electrolysis for hydrogen production for fuel cell cars and storage. This new functionality could be charged and discharged like conventional batteries and would produce hydrogen when fully charged (Table 3.2).

The advantage of nickel-iron batteries is that they do not have the lead or cadmium of lead-acid and nickel-cadmium batteries, which must be treated as hazardous materials. But the challenge concerns more about nickel supply and all its related issues. While the cost of materials is low, they are still more expensive then lead-acid batteries. NiFe battery is physically very hard to damage and has high resistance to electrical abuse, vibrations and high temperatures. However, performance is poor at low temperatures. While the battery self-discharges rapidly, it suffers no damage from full discharge. Generally, they have long life in cycling modes and standby application. Main drawbacks are low poor power density and energy density, and heavy weight.

Commonly, NiFe batteries have different properties depending on the purity of the materials. NiFe batteries with low purity will live for about 20 years, which corresponds to approximately 3000 cycles at 80% DOD and can withstand temperatures ranged from -20 to 40 °C [7]. High purity NiFe batteries will have an expected life of 30–85 years, which corresponds to approximately 11,000 cycles at 80% DOD and can withstand temperatures ranged from -30 to 60 °C. Any NiFe battery will self-discharge 1% per day at an average of 25 °C. Extreme temperatures will have repercussions on the cells, causing them to work from their usual 80% efficiency to 30%. The specific energy of the battery is low at 55 Wh/kg due to their heavy materials, but the energy density is better at 110 Wh/L.

Environmental Impact

NiFe battery is less dangerous than Lead or Cadmium based batteries, which are treated as hazardous materials. Iron is non-toxic and its extraction processes are not too costly and polluting. Nickel on the other hand is slightly toxic and harder to find. The electrolyte does not involve in the reaction and will therefore suffer no degradation. Therefore, if the battery is not correctly sealed, spills could be a problem due to the high alkaline content. Good ventilation must also be ensured in the battery surroundings, as the production of Hydrogen can potentially cause a fire hazard.

Nickel-Cadmium (NiCd) Batteries

General Characteristics

The nickel-cadmium battery is also composed of a positive nickel (III) oxide-hydroxide plates but with a negative metallic cadmium plate, and the same potassium hydroxide electrolyte. Compared to other types of rechargeable batteries, they offer good life and performance at low temperatures with average capacity, but their significant advantage is the ability to provide nearly full rated capacity at high discharge rates (can discharge in one hour or less). However, the materials are more expensive than those of the lead-acid battery and the cells have high self-discharge rates. In addition, nickel-metal hydride and lithium-ion batteries have recently become commercially available and less expensive, with the former type now rivalling NiCd batteries in cost. Where energy density is important, NiCd batteries are at a disadvantage compared to nickel-metal hydride and lithium-ion batteries.

Table 3.3 Characteristics of NiCd battery

Specific energy	40–60 Wh/kg
Energy density	50–150 Wh/L
Specific power	150 W/kg
Efficiency (charge/discharge)	<70–90%
Self-discharge rate:	10–20%/month
Operating temperature (C)	−20 to 45 °C (max 50 °C, min −30 °C)
Useful life	1500–2000 cycles
Cost	n/a Wh/US $
Applications	Sales forbidden in EU with some exceptions (medical, alarm system, emergency lightening)

NiCd batteries were once widely used in portable power tools, photographic equipment, flashlights, emergency lighting, radio controls, and portable electronic devices. The superior capacity of nickel-metal hydride batteries and their recent lower cost have largely affected the use of NiCd (Table 3.3).

Chemistry of Battery

Anode $(-)$: $2NiO(OH) + 2H_2O + 2e^- \leftrightarrow 2Ni(OH)_2 + 2OH^-$
Cathode $(+)$: $Cd + 2OH^- \leftrightarrow Cd(OH)_2 + 2e^-$

Environmental Impact

The environmental impact of cadmium (toxic heavy metal) has contributed significantly to the reduction in their use. In the European Union, since 2006 NiCd batteries have been forbidden, and can only be supplied for replacement purposes or for certain types of new equipment such as medical devices, emergency lighting, and alarm systems. Larger vented wet cell NiCd batteries are used in backup power, uninterruptible power supplies and other applications. There is ~6% of Cadmium in industrial batteries, and 18% in commercialised ones. Cadmium needs to be collected by the manufacturer in order to be recycled.

3.5.3 Lithium-Ion Batteries

General Characteristics

Lithium ion batteries make use of an intercalated lithium compound as the electrode material. These batteries are amongst the most common batteries used for domestic applications, due to their high energy density, small memory effect and low self-discharge. Lithium ion batteries are widely seen as the replacement to lead acid batteries as they can provide the same voltage outputs as lead whilst being much lighter.

Lithium ion batteries are generally composed of a negative electrode made of carbon, a positive electrode made of a metal oxide, and an electrolyte of a lithium salt in an organic solvent. There are a number of metal oxides ($LiCoO_2$, $LiFePO_4$, $LiMn_2O_4$, $LiMnO_3$, $LiNiMnCoO_2$) and electrolytes ($LiPF_6$, $LiAsF_6$, $LiClO_4$, $LiBF_4$, $LiCF_3SO_3$) which can be employed in Lithium ions batteries. The most common negative electrode used commercially is graphite. The positive electrode is normally either a layered oxide (such as lithium cobalt oxide), a polyanion (such as lithium iron phosphate), or a spinel (such as lithium manganese oxide). More recently graphene-containing electrodes (based on 2D and 3D graphene structures) have been used as electrode components for lithium batteries. The electrolyte is typically a mixture of organic carbonates containing lithium-ion complexes; non-aqueous electrolytes typically use anions salts.

Depending on material choices, the voltage, energy density, life span and safety of a lithium ion battery can change dramatically. Current efforts are focused on the use of new architectures using nanotechnology to improve performance. Areas of interest include nanoscale electrode materials and alternative electrode structures. Pure lithium is highly reactive. It reacts vigorously with water to form lithium hydroxide (LiOH) and hydrogen gas. Therefore, a non-aqueous electrolyte is generally used, and a sealed container rigidly excludes moisture from the battery pack.

Lithium-ion batteries are more expensive lead-acid batteries, but they operate over a wider temperature range with higher energy densities. Thus, they are providing lightweight, energy-dense power sources for a large variety of devices and applications. Lithium ion batteries are seen as the most promising technology for hybrid and electric vehicles [18]. In addition to their common use for cell phones now and household energy storage, this technology is currently flooding the market. One of the most popular developments of the Lithium ion technologies has been the Tesla Powerwall (Table 3.4).

Chemistry of Battery

Anode ($-$): $6C + Li^+ + e^- \leftrightarrow LiC_6$
Cathode ($+$): $LiXO_2 \leftrightarrow Li_{0.5}XO_2 + Li^+ + e^-$

Technical Data

Lithium ion batteries have an impressive specific energy; it has an average of 150 Wh/kg. This value is much higher than that of Lead-acid and Nickel Cadmium. The energy densities of LIBs average at 400 Wh/L, similar to alkaline batteries. Their temperature range go from -20 to 60 °C, making them suitable for hotter environments. The roundtrip efficiency for LIBs is approximately 85% efficiency. Currently, LIBs can be expected to cost between $350 and $700/kWh and this price is predicted to fall as the technology grows. Laboratory experiments have found that if Lithium ion batteries are cared for and kept within its safe discharge rate, they could last from 2000 cycles at 100% DOD to 5000 cycles at 65% DOD.

Table 3.4 Characteristics of Li-ion battery

Specific energy	100–265 Wh/kg
Energy density	100–600 Wh/L
Specific power	200–600 W/kg
Efficiency (charge/discharge)	80–95%
Self-discharge rate:	0.35–2.5%/month
Operating temperature (C)	−20 to 50 °C (max 60 °C, min −30 °C)
Useful life	400–6000 cycles ≈ 2–10+ years
Cost	1–6.4 Wh/US $
Applications	Calculators
	Medical equipment and pagers
	Communications equipment
	Portable radios/TVs/Cameras/Camcorders
	Emergency Power Backup or Uninterruptible Power Supply
	Dependable Electric and Recreational Vehicle Power
	Reliable and Lightweight Marine Performance
	Solar Power Storage
	Surveillance or Alarm Systems in Remote Locations

Environmental Impact
_____

The lithium-ion batteries have an environmental impact, with the biggest risk to the environment stemming from its production phase. Lithium is usually found in compound form due to its reactive nature and processing this material is detrimental to the environment. Another factor is the sourcing of this material; as Lithium is most abundant in salt flats, most of it is mined there. Mining in the salt flats requires large amounts of water, which leads to water eutrophication, transport emissions, and lack of water for local populations and wildlife. When leaching, toxic chemicals are also used. This requires waste treatment and there is a concern of it not being disposed of correctly. When lithium-ion batteries have reached their useful life, there are limited ways in which they can be disposed of, and most batteries end up in a landfill.

Lithium-Iron Phosphate Batteries

$LiFePO_4$ is considered one of the best Lithium ion chemistries for high performance accumulators. Whilst their specific energy is not extremely high at 110 kWh/kg, their specific power output is, at 200 W/kg. The iron phosphate chemistry is superior to others such as cobalt oxide and manganese-spinel due to its greater abundance and improved thermal performance. Due to iron phosphate lower internal resistance, thermal runaway is less likely to occur under heavy charge/discharge cycling. Although they have better thermal management, careful consideration must still be given to the cooling of the cells, which are predominately cooled with corrugated aluminium heat sinks and consistent airflow. At full charge, they can be stored to up to a year. Their expected life can be from 500 at 80% DOD, to 5000 cycles at 50% DOD.

LiFePO4 battery is non-toxic as it does not contain any heavy or rare materials. It also complies with the European RoHS (Restriction of Hazardous Substances). Majority of the CO_2 emissions are production related. During its useful life, it produces less than half the output from its production stage. The emissions change with the different charge cycle rate; the highest cycle rate batteries are the smallest, as they need to be working at an increased rate to keep up with the required power.

Sodium Ion Batteries

Sodium ion batteries are seen to be an answer to Lithium ion batteries as the operate very similarly. Sodium ion batteries contain Sodium compounds as the cathode of the battery. The anode is predominantly Carbon based. The electrolytes usually have a range of materials used to suit the specific Sodium compound that is used; these will range from salts, ether solvents and polymer electrolytes. Where Lithium ion batteries have a limited availability, making the material expensive to procure, Sodium as the material is almost unlimited as it makes up a majority of the Earth's crust [1].

Sodium batteries can have a ranging cycle life; depending on the type of compounds that are used, the batteries can have a life exceeding 300 cycles. Articles have listed a range of 1000 cycles to 5000 cycles whilst still retaining 80% of capacity.

The best qualities of Sodium ion batteries are the discharge depth and the roundtrip efficiency, although charge and discharges are slow. The DOD can theoretically be 100% discharge; however, this will not be the case if the battery as it is most desirable for it to have a greater life span. The temperature range for the battery is around the average for most battery materials. This is suitable for most grid uses, assuming there is no extreme weather. The specific energy of the Sodium is stated to range from 50 to 102 Wh/kg, making them quite heavy [10, 15]. The energy density for Sodium ion is stated to be even higher than LIBs with values ranging from 600 to 650 Wh/L [4] and the available capacity stands at 13.5 kWh with a 4h discharge time [6].

The price of Sodium as a source material is one of the factors as to why this battery is so popular. The materials are extremely abundant and therefore the cost of Sodium ion batteries is much lower than some other battery types. It has been estimated that the price of Sodium ion batteries is around 30% cheaper than Lithium ion batteries [2]. The overall cost/kWh is similar to that of a lead battery and can be estimated to be 200–250$/kWh.

Sodium ion batteries are suitable for off grid storage systems (domestic solar storage capabilities), grid storage, large scale renewable energy storage (wind and solar) and electric or hybrid cars applications. Safe storing and shipping in addition to desirable technical properties and low cost make them an attractive option.

Environmental Impact

The global impact of the Sodium ion battery is linked to its global warming potential and in the marine eutrophication potential. If sugar beet is farmed for the production of Carbon for the anode, the environmental issues arise due to its contribution to global warming. The disposal of this product is detrimental to marine eutrophication. In regards to the cathode itself where the Sodium will be, the main factor to the environment comes from terrestrial acidification, as it can lower the pH in the soil in a detrimental manner.

3.5.4 Flow Zinc-Bromine Batteries

General Characteristics

The flow batteries are a type of rechargeable battery where the two chemical components of its chemical reaction are in liquid state and are pumped through the system on separate sides of a membrane. Ion exchange occurs across the membrane while the two liquids flow in their own respective spaces. Although they have technical advantages over conventional rechargeable batteries, such as potentially separable liquid tanks and virtually unlimited longevity, current implementations are comparatively less powerful and require more sophisticated electronics. In general, the energy capacity depends on the electrolyte volume, and battery power on the electrode area.

Different from others, the Zinc-Bromine (ZnBr) battery is a hybrid flow battery. A Zinc bromide electrolyte solution is stored in two different tanks and pumped between them: one tank stores the electrolyte for the positive electrode reactions and the other for the negative. This circulating motion will keep the reactant supply uniform and equal to each cell. During charging, metallic Zinc is plated from the electrolyte solution onto the negative electrode surfaces in the cell batteries. The bromide is converted to bromine at the positive electrode surface and is stored in a safe, chemically complexed organic phase in the electrolyte reservoir. During discharge, the reverse process occurs: the zinc metal plated on the negative electrodes dissolves in the electrolyte and is available to be plated again in the next charge cycle. It can be left fully discharged indefinitely without damage.

Zinc-bromine batteries have high specific energy compared to lead acid batteries. Further advantaged include 100% depth of discharge capacity on a daily basis and no shelf-life limitation, as zinc-bromine batteries are not perishable, unlike lead-acid and lithium-ion batterie. The technology is scalable. However, there are several disadvantages: the battery needs to be fully discharged every few days to avoid zinc dendrites that can puncture the separator, and every 1–4 cycles to short the terminals on a low impedance shunt while running the electrolyte pump, in order to completely remove zinc from the battery plates. Low power density during charging and discharging results in high energy cost.

This type of battery is of particular interest for remote telecom sites application, where significant fuel savings are possible when they are operating under low electrical load and high installed generation conditions by using multiple systems in parallel to maximize the benefits and minimize the drawbacks of the technology.

Table 3.5 Characteristics of ZnBr battery

Specific energy	65–75 Wh/kg
Energy density	60–70 Wh/L
Specific power	100 W/kg
Efficiency (charge/discharge)	70–80%
Self-discharge rate:	20–30%/month
Operating temperature (C)	15–50 °C (max 55 °C, min 0 °C)
Useful life	No cycle limitation ≈ up to 50 years
Cost	0.4 Wh/US $
Applications	Telecommunication sites Power solar systems Power wind systems Other off grid applications

Nevertheless, they still compete to provide energy storage solutions at a lower overall cost than other energy storage systems such as lead-acid, lithium-ion and others (Table 3.5).

Chemistry of Battery

Anode (−): $Zn^{2+} + 2\,e^- \leftrightarrow Zn_{(s)}$
Cathode (+): $2Br^-_{(aq)} \leftrightarrow Br_{2(aq)} + 2\,e^-$

Technical Data

One of the biggest advantages of ZnBr batteries is that the battery performance is not sensitive to cycle depth; there are no cycle limitations. This makes the battery one of the top market choices as technically it could easily last over 50 years in any application. It has a 100% depth of discharge with no potential damage and an efficiency which varies from 70% to 80%. However the recommended temperatures range from 15 to 50 °C which is a small range. The battery work up to a range of 0–55 °C, but no more. One of its strong points is that it has a high peak energy density of 10 kWh. Specific energy of the battery ranges from 65 to 75 kWh/kg and its energy density per litre ranges from 60 to 70 Wh/L. While energy density is high, power output is low, and battery can be heavy and large. Battery life is impressive and it is not flammable.

Environmental Impact

Zinc is a common metal which is available and relatively inexpensive. It can be manufactured easily and is not a risk towards the environment. Bromine on the other hand is toxic and as Lead, should be treated as a hazardous material. The filtration of Bromine would potentially have a high negative impact on the environment if filtrated into water or soil. This battery is fully recyclable.

3.6   Comparison of Battery Technologies

With numerous relevant technical parameters to consider when choosing an energy storage solution, and many technologies available, comparison graphs are useful for quick evaluations of batterie types with desirable technical characteristics.

Comparison of the specific energy and the specific density is essential for battery selection, showed in Fig. 3.6. Higher specific energy means lighter battery; higher specific density means smaller battery size. In terms of performance alone, lithium-ion technologies are clearly dominant. Nevertheless, some technologies are also of interest, especially the lithium-ion polymer batteries, the sodium sulfate batteries or the sodium-ion batteries.

The weight disadvantage is present for Nickel metal batteries, which include NiFe; remain at approximately at the lower half of the y-axis which corresponds to their Watthours/kg value. Nevertheless, specific density is much better than other conventional batteries, comparable to some lithium batteries.

The battery performance, with respect to its power and energy density, is the main consideration for mobile applications such as an electric car, where increasing either the maximum power output or capacity leads to an increase battery weight. Therefore, the optimal battery for a mobile application (e.g. electric car) results from selecting a battery type with a balance of both power and energy density. In between, rechargeable batteries can fit with applications requiring a discharging time from around some minutes to some days, where thermal storage can go up to some months. But since the energetic density of batteries is a way much higher than that of thermal storage, this makes them a much more practical technology for small or mobile applications, where volume and weight are critical parameters.

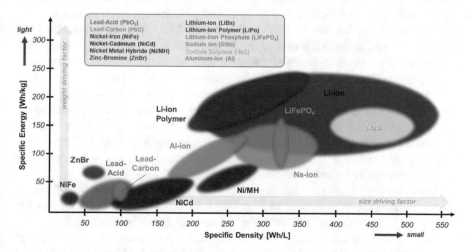

Fig. 3.6 Specific energy and specific density comparison for different chemical batteries

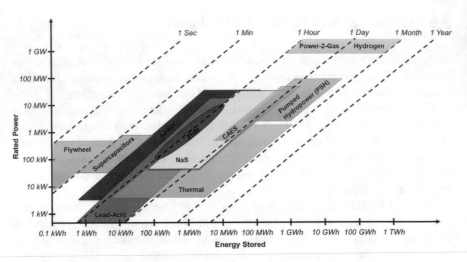

Fig. 3.7 Rated Power and Stored Energy comparison for different battery chemistries, amended from [13]

In Fig. 3.7, we are comparing the rated power and the energy stored for battery technologies and other types of storage systems in terms of their discharging time range. Flywheel technology or supercapacitors are suitable for a quite short discharging times and can store relatively small amount of energy, whereas other storage solutions like Power-2-Gas or Hydrogen are better for long-term energy storage with high rated power and energy stored capacity.

The potential for use in grid storage applications for molten salt batteries with a Sodium Sulphur (NaS) chemistry. This is due to their high energy density (150 Wh/kg), roundtrip efficiency of 75%, and a lifetime of 2500 cycles at 100% depth of discharge (DOD), or 4500 cycles at 80% DOD. NaS batteries have been successfully scaled up to the tens of megawatt hours, while providing a high rated power compared to other battery chemistries. NaS batteries are suited for large scale energy grid applications due to their inexpensive and abundant materials. Due to the simplicity of the cells and the abundance of required raw materials, they become more economical to manufacture the larger they are scaled up. When considering the performance of NaS batteries they display; reasonable power and energy densities as well as temperature stability. The next step to making these batteries more commercially viable is to lower their operating temperature via research of new chemistries such as ceramic and glass electrolytes [17].

Investment cost is an essential parameter because it determines the overall cost of energy utilisation and therefore influences the final cost to the consumer. Investment cost per energy unit and the cost per power unit for 4 batteries technologies, 3 different mechanical storage technologies and supercapacitors is presented in Fig. 3.8. The values presented are rough estimates of an order of magnitude to allows for the comparison of different types of technologies. The effective investment cost of each technologies is very hard to determine accurately because it depends on a

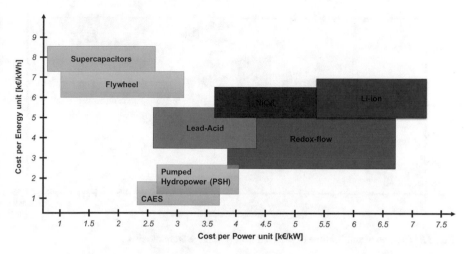

Fig. 3.8 Investment cost for selected energy storage technologies

large number of parameters (including specific application, manufacturer, market, place and time).

Compressed Air Energy Storage and Pumped Hydropower technology have the lowest investment cost per unit of power and energy. Hence, 96% of the total grid-connected operational stored capacity worldwide are based on this technology. Flywheel and supercapacitors have lower cost per power unit since they are designed to deliver energy at high power rate. However, in terms of energy cost, since they are storing large amount of energy, their investment cost per energy unit is the highest.

Lead-acid battery technology has the lost investment cost, which explains why this technology have been utilised widely for transport applications throughout history. For solar application nickel-cadmium, lithium-ion and redox-flow batteries hare of interest, even though they have a higher capital cost. In addition to the investment and maintenance costs over the lifetime of the project, the expected life span of the battery technology must not be overlooked.

In Fig. 3.9 efficiency and the lifetime of different storage technologies are shows. An ideal storage system would have the highest efficiency with the longest lifetime or number of cycles. Only supercapacitors and flywheels have these two properties, but are already mentioned, they are limited in term of stored energy capacity. Lithium-ion batteries have the best efficiency and life span compared to other battery technologies. It would be interesting then to compare different battery technologies for the storage part of a PV system over its whole lifetime, since lithium-ion batteries are certainly more expensive than lead-acid for example, but have a higher expected number of cycles. Then that would help to choose the most suitable technology, at least from the financial point of view.

Figure 3.10 illustrates the maturity different storage technology using the Technology Readiness Level (TRL) parameter. This is an illustrative example to compare different technologies, since their research and development is progressing rapidly

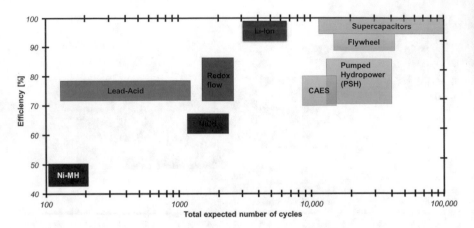

Fig. 3.9 Efficiency and lifetime of selected energy storage technologies

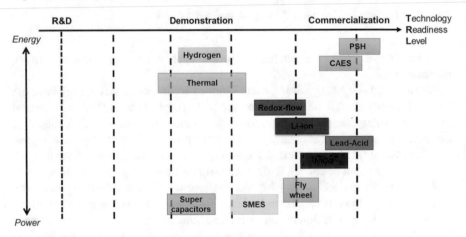

Fig. 3.10 TRL for selected energy storage technologies

and their TRL is always changing (TRL can be a subjective parameter). Pumped Hydropower and Lead-acid batteries are the most mature technologies since they have been used for several decades, closely followed by Compressed Air Storage, Nickel-Cadmium batteries. Li-ion batteries have been growing very quickly over the last decade. Feasibility of flow batteries and flywheels have been demonstrated but these technologies are still struggling to get established in a sustainable way worldwide. Regarding Hydrogen, supercapacitors and thermal storage, there are have several operational industrial pilot projects underway, with expectations to facilitate their development and commercialisation in the coming years.

3.7   Summary of Battery Characteristics (Table 3.6)

Table 3.6 Comparison of battery characteristics

	ZnBr	Li-ion	NiFe	Lead acid
Specific energy (Wh/kg)	65–75	100–265	19–25	35–40
Energy density (Wh/L)	60–70	100–600	30	80–90
Specific power (W/kg)	100	200–600	100	140–180
Efficiency (charge/discharge)	70–80%	80–95%	<65%	50%/95%
Self-discharge rate:	20–30%/month	0.35–2.5%/month	20–30%/month	3–20%/month
Operating temperature (C)	15 to 50 °C	−20 to 50 °C	−20 to 40 °C	−10 to 40 °C
Useful life	No cycle limitation Up to 50 years	400–6000 cycles 2–10+ years	3000–11,000 cycles 20–85 years	500–1000 cycles 4–5 years
Applications	Solar power systems Wind power systems Other off grid applications	Emergency Power Backup or Uninterruptible Power Supply Dependable Electric and Recreational Vehicle Power Reliable and Lightweight Marine Performance Solar Power Storage	Wind and solar power systems where weight is not important Off-grid applications Daily charging applications Hydrogen production Fuel cell cars and storage	Automotive and traction application High current drain applications Sealed battery types available for use in portable equipment Grid scale energy storage

3.8   Battery Management Systems

A Battery Management System (or BMS) is an embedded system made up of purpose-built electronics and software intended to perform the appropriate management of a battery pack. The hardware elements of the BMS incorporate electronic circuits to ensure the safety of the battery pack and its operator and make measurements that include voltages, electrical current, and temperature. The software part of the BMS monitors and coordinates the activities of the battery pack.

A battery module is any number of identical battery cells that may be connected in series, parallel, or both to deliver the desired voltage, capacity, or power density.

A battery pack is a set of any number of identical battery modules. They may be connected in series, parallel, or in a mixed configuration, to deliver the desired voltage, capacity, or power density.

Battery management systems are not used in simple battery-powered systems with low replacement costs and no safety issues. Instead, battery management systems are used in mission-critical or large battery installations, where replacement costs are high, or there may be safety issues. BMS-equipped devices include electric vehicles, smart residential batteries, or utility-scale battery storage systems. The primary purposes of a battery-management system are:

1. to ensure the safety of the battery-powered system's operator. The BMS must be able to identify and respond to hazardous operating conditions. This requirement may necessitate isolating and separating the battery pack from the load and informing the operator.
2. to safeguard the battery pack's cells from damage in the event of misuse or failure. This requirement could include software-controlled active intervention or specialised electronics that can detect faults and disconnect the malfunctioning components from the rest of the battery pack and the load.
3. to extend the battery pack's useful life in normal operating conditions. The BMS accomplishes this by coordinating with the load controller and notifying it of power restrictions that must be respected over a short time interval, ensuring that the battery pack is not overcharged or over drained. The BMS also regulates the thermal-management system, ensuring that the battery pack remains within its specified operating temperature range.
4. to keep the battery pack in a condition that allows it to meet its functional design criteria.

The functional requirements of a battery management system can be categorised as follows:

1. *Sensing and control:* the BMS must monitor cell voltages, temperatures, and battery pack current. It must also detect isolation failures, control the contactors or electronic switches that protect the battery pack and the thermal management system by operating devices such as a pump, fans, or a heater.
2. *Battery pack protection:* the battery management system must incorporate electronics and logic to safeguard the operator and the battery pack itself from overcharging, over-discharging, overcurrent, cell short circuits, and high temperatures. The protection measures can be physical (emergency disconnection of the battery) or informational (reporting problems to the user).
3. *Information Interface:* where possible, the BMS must interact with the system powered by the battery pack frequently, reporting available energy, power, and other battery-pack status indications. Alternatively, the BMS may provide relevant information about the battery-pack status through an electronic display. Some systems also keep a permanent record of uncommon error or abuse events for diagnosis purposes.

4. *Battery pack performance management:* the battery management system must estimate the state of charge (SOC) for all cells in the battery pack, balance the cells, and compute available energy and power limits for the battery pack.

5. *Diagnostics:* The battery management system must be able to estimate quantitative parameters that correlate with the health of the battery pack, including the degradation of battery capacity and the increase of internal resistance, both of which generally occur over the lifetime of the battery pack. The system should also be able to detect misuse of the battery pack.

References

1. Adelhelm, P., Hartmann, P., Bender, C. L., Busche, M., Eufinger, C., & Janek, J. (2015, April). From lithium to sodium: cell chemistry of room temperature sodium–air and sodium–sulfur batteries. *Materials for sustainable energy production, storage, and conversion, 6.* Retrieved February 2017, from http://www.beilstein-journals.org/bjnano/single/articleFullText.htm?publicId=2190-4286-6-105

2. Casey, T. (2016, March 14). *Faradion Leads Energy Storage Trio Into Sodium-Ion Territory.* Retrieved February 2017, from Clean Technica: https://cleantechnica.com/2016/03/14/solar-energy-storage-targeted-by-sodium-ion-trio/

3. Chen, H., Cong, T. N., Yang, W., Tan, C., Li, Y., & Ding, Y. (2009). Progress in Electrical Energy Storage System: A Critical Review. *Progress in Natural Science,* 291–312. https://doi.org/10.1016/j.pnsc.2008.07.014

4. Choi, J., Aurbach, D. Promise and reality of post-lithium-ion batteries with high energy densities. *Nat Rev Mater* 1, 16013 (2016). https://doi.org/10.1038/natrevmats.2016.13

5. Cook, T. R., Dogutan, D. K., Reece, S. Y., Surendranath, Y., Teets, T. S., & Nocera, D. G. (2010). Solar Energy Supply and Storage for the Legacy and Nonlegacy Worlds. *Chem. Rev.,* 6474–650. https://doi.org/10.1021/cr100246c

6. European Association for Storage of Energy. (2016). Sodium-Ion Battery: Electrochemical Energy Storage. *Energy Storage Technology Descriptions,* 1–2. Retrieved from http://ease-storage.eu/wp-content/uploads/2016/07/EASE_TD_Electrochemical_NaIon.pdf

7. Gaffor, S. A., Haripraksh, B., & Shukla, A. K. (2010, July). Nickel-iron battery-based electrochemical energy storage systems for rural/remote area telecommunication. *Telecommunications Energy Conference (INTELEC), 32nd International,* 1. Retrieved from http://ieeexplore.ieee.org/document/5525702/

8. Khan, N, Dilshad, S, Khalid, R, Kalair, AR, Abas, N. Review of energy storage and transportation of energy. *Energy Storage.* 2019;e49. https://doi.org/10.1002/est2.49

9. Linden, D., & Reddy, T. B. (2001). *Handbook of Batteries (3rd edition).* New York: McGraw-Hill Companies Inc.

10. Peters, J., Buchholz, D., Passerini, S., & Weil, M. (2016, March). Life cycle assessment of sodium-ion batteries. *Energy & Environmental Science,* 2–5. https://doi.org/10.1039/C6EE00640J

11. Posada, J. O., Abdalla, A. H., Oseghale, C. I., & Hall, P. J. (2016). *Multiple regression analysis in the development of NiFe cells as energy storage solutions for intermittent power sources such as wind or solar.* International Journal of Hydrogen Energy, Volume 41, Issue 37, 5 October 2016, Pages 16330–16337, https://doi.org/10.1016/j.ijhydene.2016.04.165

12. Rydh, C. J. (1999). Environmental Assessment of Vanadium Redox and Lead-acid Batteries for Stationary Energy Source. *Journal of Power Sources 80,* 21–29. https://doi.org/10.1016/S0378-7753(98)00249-3

13. Sapub. (2016, August 6). *Rated Power VS Energy of Various Battery Types.* Retrieved from Sapub: http://article.sapub.org/image/10.5923.j.eee.20160601.01_006.gif

14. Shen, P. K., Wang, C.-Y., Jiang, S. P., Sun, X., & Zhang, J. (2016). *Electrochemical Energy: Advanced Materials and Technologies.* Published July 27, 2017 by CRC Press ISBN 9781138748927
15. Shirpour, M., Zhan, X., & Doeff, M. M. (2015, September). Sodium-Ion Batteries: "Beyond Lithium-Ion". *Conference: 2015 TechConnect World Innovation Conference (DC)*
16. Svarc, J. (2021, September 30). *Complete Home Battery Guide 2021.* Retrieved from Cleanenergyreviews: https://www.cleanenergyreviews.info/blog/home-solar-battery-cost-guide
17. Wen, Z., Cao, J., Gu, Z., Xu, X., Zhang, F., & Lin, Z. (2008). Research On Sodium Sulfur Battery For Energy Storage. *Solid State Ionics 179*, 1697–1701. DOI:https://doi.org/10.1016/j.ssi.2008.01.070
18. Zackrisson, M., Avellán, L., & Orlenius, J. (2010). Life cycle assessment of lithium-ion batteries for plug-in hybrid electric vehicles – Critical issues. *Journal of Cleaner Production*, Volume 18, Issue 15, November 2010, Pages 1519–1529 https://doi.org/10.1016/j.jclepro.2010.06.004

Chapter 4
Concentrating Photovoltaics

4.1 Introduction

Concentrating Photovoltaics (CPV) is a technology that associates a concentrator with a photovoltaic device as shown in the Fig. 4.1. In a more detailed way, the concentrator is actually one or a series of optical devices that concentrate the sun beams onto a solar cell in order to increase the electrical output of the photovoltaic device by increasing the intensity of input incident solar power. But, why should one use concentrating photovoltaics instead of normal photovoltaics? Since CPV uses optical devices to concentrate sunlight onto the solar cells, this allows for a reduction in the cell area required for producing a given amount of power. Thus, the global aim of this technology is to significantly reduce the cost of electricity generated by replacing the expensive cell area with less expensive optical material (Fig. 4.2). This approach also provides the opportunity to use higher performance multi-junction cells that would be prohibitively expensive without concentration. As a result, CPV modules can easily exceed 30% energy conversion efficiency (they are expected to exceed 50% in the medium term), whereas the common PV technology has a maximum efficiency of about 25%. Therefore, CPV technology can be seen as an innovative and enhanced version of the usual photovoltaics that would produce more electrical power output at lower cost. Before going a bit further into the details of this technology, it is important to have in mind what exactly is concentration.

4.2 Concentration

Concentration is the ability of a concentrator or an optic system to concentrate solar radiation. Then, in order to compare all the different concentrators, a concentration ratio has been defined as the quantity of solar radiation reaching the receiver compared to the quantity of solar radiation entering the aperture area of the

A. Rachid et al., *Solar Energy Engineering and Applications*, Power Systems, https://doi.org/10.1007/978-3-031-20830-0_4

Concentrator   Photovoltaics

Optical device   Solar cells

Fig. 4.1 Concentrator and the photovoltaics

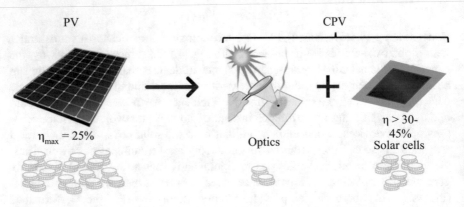

Fig. 4.2 Replacement of expensive cell area with less expensive optical material

concentrator as shown in the Fig. 4.3. Thus, the higher the gap between these two values of solar flux, the higher the concentration ratio. From a more geometrical point of view, the geometrical concentration ratio is the ratio of the aperture area to the receiver area which can be increased by increasing the aperture area and decreasing the receiver area.

A concentration ratio scale has been defined in order to classify the various CPV technologies in terms of their capacity to concentrate solar radiation onto a solar cell as shown in the Fig. 4.4. It starts from the value 1 (which corresponds to no concentration) up to several thousand. With this scale, it has been considered that:

- low concentration technologies are between 1 and 10 suns,
- medium concentration technologies are between 10 and 100 suns (which sometimes are also assimilated to low concentration)
- high concentration technologies are between 100 and 2000 suns,
- and ultrahigh concentration technologies are above 2000 suns.

Fig. 4.3 Concentration ratio

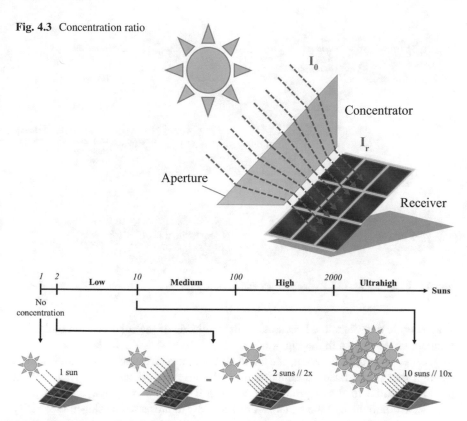

Fig. 4.4 Concentration ratio scale

Nevertheless, there are limits on the concentration ratio up to which one can reach. For example, the maximum concentration ratio that can be reached for 2D optics is around 216 suns because of the fact that the sun is not a point source which makes the sun beams slightly unparallel to each other. The same reason limits the maximum concentration ratio to 46 000 suns for 3D optics. Other limitations exist that can be due to the type of optics or the geometry of the concentrator. To summarize the concentration ratio is a key parameter for characterising and classifying CPV technologies. However, it is not the only one. There are several ways to classify the CPV technologies. However, they all remain based on optics, but it can be in function of their geometry, the type of their primary or secondary optics, or about their tracking method as shown in the Fig. 4.5. Nevertheless, the classification based on the concentration ratio remains the most usual and convenient.

Now that some key definitions related to CPV systems have been discussed, the focus will now be on the main components of this technology, before looking at the different options that can be considered for each of them. From a global point of view, a CPV basic unit is composed of a solar cell, a concentrator that can be one or

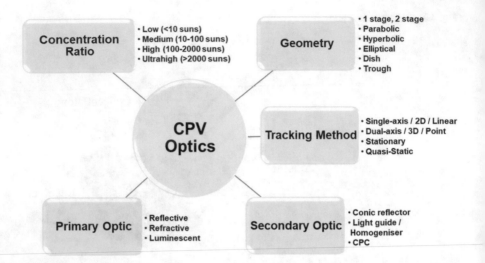

Fig. 4.5 Classification of CPV technology and its optics [1]

several optic devices and a heat sink on the backside of the solar cell to regulate its temperature as shown in the Fig. 4.6.

Indeed, with concentration, the incident solar flux arriving on the solar cell is so high that it can raise its temperature and cause it to deteriorate. Then, a heat sink is essential to firstly protect the solar cell – which is by the way the most expensive component of the system – and secondly to preserve its solar-to-electricity conversion performance that drops drastically when its temperature raised. As a matter of fact, an increase in solar cell temperature of approximately 1 °C causes an efficiency decrease of about 0.45%.

The basic units of CPV can be connected together to form a panel as shown in Fig. 4.7. Similarly, several panels can also be connected to form a module (Fig. 4.8) and several modules can be connected together to form a whole system that can be placed onto a tracking device, the last component of CPV technology as shown in Fig. 4.9.

Several systems can be placed on the ground and connected together to form a CPV farm. The components required to transport, transform or store electricity are roughly the same as for the usual PV farm and thus they are not discussed in this chapter.

As the four main components of the CPV technology have just been listed, more details on each of them will now be given in the subsequent sections starting with the most important component which convert solar radiation into electricity, the solar cells.

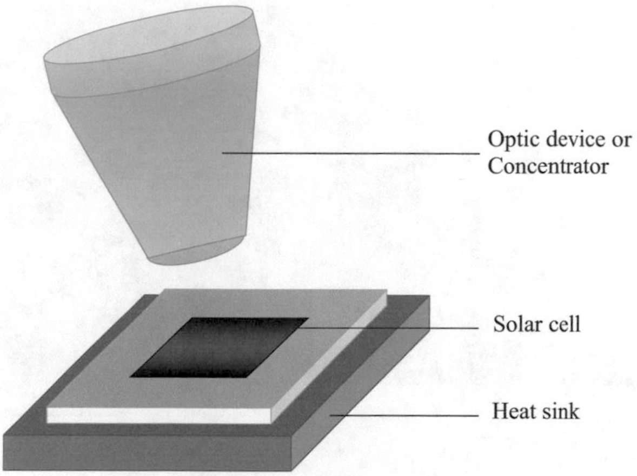

Fig. 4.6 Basic unit of CPV
technology

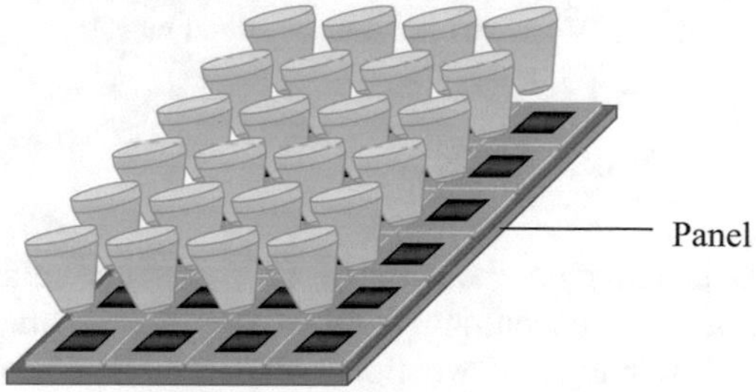

Optic device or
Concentrator

Solar cell

Heat sink

Fig. 4.7 CPV panel
consisting of several basic
units

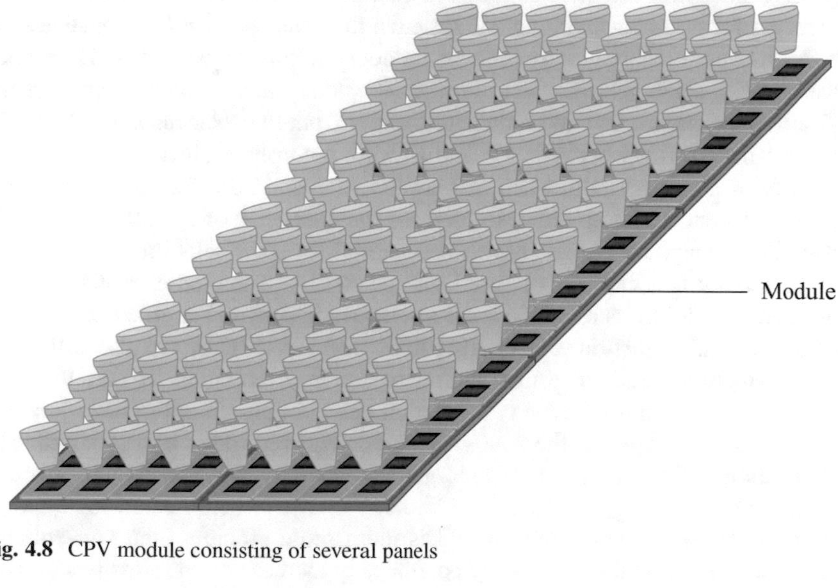

Panel

Module

Fig. 4.8 CPV module consisting of several panels

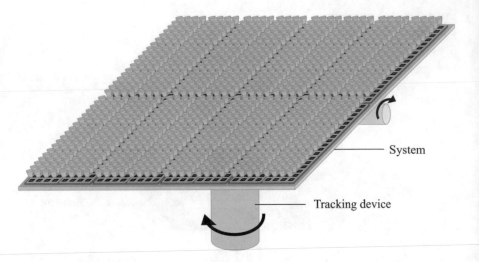

Fig. 4.9 CPV system consisting of several modules

4.3   Solar Cells

Depending on the type of CPV system, solar cells technologies are different. For low concentrating PV, where it is a question of increasing the efficiency at lower cost, conventional silicon cells are normally used. While the high and medium concentration use multi-junction cells that are more efficient but also more expensive since they are more complex to manufacture and use rarer materials than silicon. As for non-concentrating PV systems, the cells used for low concentration are the conventional polycrystalline or monocrystalline silicon cells. Developed since the 70s, these solar cells are now a completely mature technology that are basically the cheapest way to produce electricity from the solar resource. The cells have an efficiency only between 8% and 13% for amorphous silicon, between 10% and 14% for a polycrystalline cell or between 11% and 20% for monocrystalline. But the low concentration allows to increase the efficiency of the cells up to 27%. Nevertheless, using silicon as converting material for solar cells limits the amount of solar energy that can be converted into electricity. Thus, other materials have been used in order to increase the solar-to-electricity conversion efficiency. During the 80s, the multi-junction cells starting to be developed. They are also called III-V multi-junction cells, since they use elements from columns III and V of the periodic table. Starting with the two-junction cell configuration, the three-junction configuration developed in the 2000s, can be considered today as a standard. The four-junction configuration is an emerging technology. With the multi-junction configuration, this type of solar cell technology allows to better catch solar energy from different wavelengths, which makes them more efficient than conventional silicon solar cells and their efficiency continues to increase as one adds junctions.

Nevertheless, the development of this type of solar cells takes time, since there are a lot of possible configurations of junctions and elements, and these cells are more expensive than conventional silicon cells. As a matter of fact, 61% of CPV systems use silicon cells, while only 17% of CPV systems employ the III-V multi-junction ones.

As a summary about solar cells, the NREL provides a comparative diagram for the evolution of different types of solar cells' efficiency with details whether they are multijunction, thin-film or silicon based solar cells and also whether they are used under concentration or not. It has been shown that the multi-junction cells are the most performant, with efficiency between 26% and 44%, where crystalline silicon cells are between 20% and 27% only. Also, it is important to have in mind that emerging PV cells have for now an efficiency only between 7% and 14%, but they are based on innovative approaches that could overcome some of the inherent issues of the multi-junction cells. For instance, organic cells use elements that are much less scarce than multi-junction ones, which can decrease the environmental impact of the production of solar cells and their cost.

4.4 Optics

Regarding the optics in CPV, there are actually two types of devices that are used. First, there are mirrors that are devices that reflect solar radiation, and secondly, there are lenses which are devices where solar radiation passes through. These two types of optics differ in their geometry, where a large number of configurations is possible. Only the most common ones have been presented here, from the simplest one like the flat reflector or the V-trough, to the more complex ones such as the Fresnel reflector or lens, the parabolic Dish or the rod lens as shown in Fig. 4.10. They are all characterised by the following parameters: reflection, refraction, luminescent, concentration ratio and acceptance angle.

In CPV systems, a single optical device can be used to concentrate solar radiation onto the solar cell, as shown in Fig. 4.10 or a combination of several optical devices can be created resulting in a more complex concentrator as shown in Fig. 4.11. Three different examples of concentrators composed of two or three optical units have been shown. The first one is a double reflector concentrator, so-called the Cassegrain mirror [2], which is composed of a concave mirror as a primary concentrator and a convex mirror as a secondary concentrator. The second one is an association of a Fresnel lens with a Rod lens and the last one here is an association of three optics, composed of a Cassegrain mirror and a Rod lens.

These are actually only three examples of existing CPV concentrators among many others that exist. In order to compare them, their global performances have been based on three main characteristics:

– Their optical efficiency, which is the ratio of the energy arriving on the concentrator aperture by the energy arriving on the solar cell.

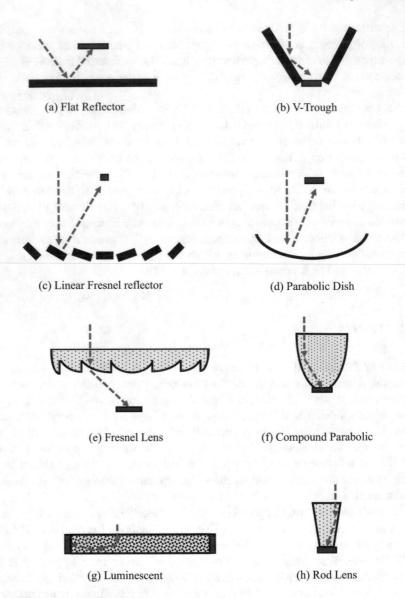

(a) Flat Reflector

(b) V-Trough

(c) Linear Fresnel reflector

(d) Parabolic Dish

(e) Fresnel Lens

(f) Compound Parabolic

(g) Luminescent

(h) Rod Lens

Fig. 4.10 Various types of optical devices

- Their acceptance angle, that corresponds to the maximum angle at which incoming sunlight can be captured by the concentrator.
- Their irradiance uniformity, that characterises the distribution of the concentred sun rays on to the solar cell.

Each CPV concentrator has its own advantages and disadvantages related to these parameters and are chosen depending on the application. Following the concentra-

Fig. 4.11 Combination of several optical devices to concentrate the sun rays

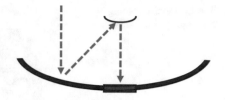

(a) Cassegrain Mirror

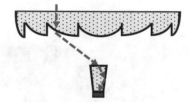

(b) Fresnel Lens + Rod Lens

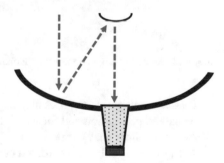

(c) Cassegrain Mirror + Rod Lens

tors, an essential component that is especially required by the CPV technology, the heat sinks have now been discussed in the subsequent section.

4.5 Heat Sinks

One of the main problems of concentrating photovoltaics is the poor efficiency of the cells when its temperature rises. As a reminder, an increase of 1 °C of the solar cell temperature will drop its efficiency by around 0.45%. Thus, a heat sink is absolutely essential to protect the most expensive component of the system i.e. the solar cell and to preserve its solar-to-electricity conversion performance. For CPV technology, two main methods of solar cells cooling are considered:

Fig. 4.12 Passive cooling of concentrating photovoltaics

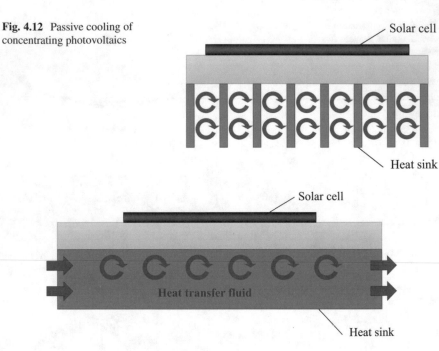

Fig. 4.13 Active cooling of concentrating photovoltaics

- Passive cooling, that requires no energy consumption, and which is mainly based on natural convection and for which heat cannot be recovered.
- Active cooling, that requires energy consumption to circulate a cooling fluid. Active cooling is based on forced convection and the heat can be recovered and used for a different application.

Passive thermal dissipation cooling systems are mostly thermal "radiators" (Fig. 4.12), inspired by the form that already exists in the field of microelectronics for cooling microprocessors engraved on silicon wafers. They are less efficient than active cooling systems but are generally cheaper and easier to implement in the CPV system. They also require less maintenance.

Active thermal dissipation cooling systems (Fig. 4.13) mostly use water, or glycol water as heat transfer fluid, from which the thermal energy can be used for hot water, heating or desalination of sea water. They are more efficient than passive cooling systems but are also more complex to implement and require more maintenance, which increases the cost of the CPV system.

4.6   Tracking Systems

There are two main types of tracking systems. The first one is the single-axis tracker where the panel rotates around one axis, generally in an azimuthal shift from east to west over the course of a day (Fig. 4.14) and the tilt angle needs manual intervention depending on the day/month. The second one is the dual-axis tracker where the panel rotates along an east-west axis and along a vertical axis as shown in Fig. 4.15. Within the single-axis tracker category, there are two further types: the active trackers, that consume energy, since they use motors, gears or hydraulics to move the module. And the passive trackers, that do not consume energy, but use a specific device that contains a compressed fluid with a low boiling point, which evaporates into a gas when heated up by the Sun, generating a mass transfer process that adjusts

Fig. 4.14 Single axis tracker

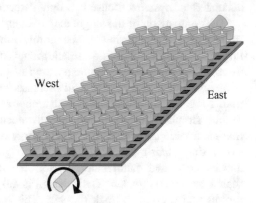

Fig. 4.15 Dual-axis tracker

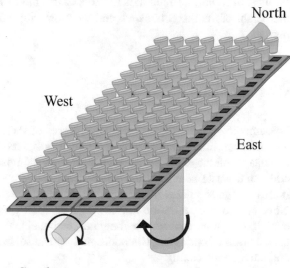

Table 4.1 Comparison of the single axis tracking and dual axis one

Single axis tracking	Dual axis tracking
Follows the sun from east to west with a single pivot point	Tracks the sun from east to west and north to south using two pivot points
Uses a predictable pattern based on the time of year to track the sun	Uses a collector "cyc" to visually track the sun
Simple and efficient design	More complex design (more motors, sensors and maintenance), but more accurate
Lower cost compared to dual axis	Higher cost due to additional parts and installation time
Few failures and malfunctions	More failures and malfunctions

the equilibrium of the CPV system. Nevertheless, these passive trackers are less accurate than systems focused on light detection technologies.

The comparison of the single axis tracking and dual axis one has been presented in Table 4.1. Considering the economics, the choice of which tracker to use is largely a decision involving the additional investment and maintenance cost over time, versus the increase in energy (and financial) output delivered by the system. Although a dual-axis tracker will increase the solar efficiency of the CPV system than a single-axis tracker, it is also important to point out that a single-axis will require simpler equipment and the manufacturers of single-axis equipment claim that the additional benefit of the net energy output provided by a dual-axis system over a single-axis system is often lost through the additional cost of its installation and its associated maintenance. Also, dual-axis systems have more moving parts than a single-axis tracker. Their design is more complex with the use of additional motors and sensors to track the sun. This increases the risk of malfunctions and failures. In addition, single-axis trackers tend to have a lower profile, sometimes half the height of dual-axis trackers, and therefore, are more likely to receive building permits.

4.7   Summary

To summarize, the strengths and weaknesses of the CPV technology considering technical and industrial aspects have been presented in the Table 4.2. The main advantage of the CPV is its high efficiency compared to PV and its possible additional waste heat recovery when using active cooling. In addition, CPV also increases and stabilizes the energy production throughout the day due to tracking. Nevertheless, concentration implies harnessing only the direct solar radiation part - even if there is some research about harnessing also the diffuse radiation part – and the addition of optics implies additional optical losses, a need of reliable tracking and a frequent cleaning.

Table 4.2 Strengths and weaknesses of the CPV technology

	Strengths	Weaknesses
Technical aspects	High efficiencies Use of waste heat possible for systems with active cooling Increased and stable energy production throughout the day due to tracking	A part of solar radiation (diffuse) cannot be used Additional optical losses Tracking required Require frequent cleaning to mitigate soiling losses
Industrial aspects	Low CapEx Modular kW to GW scale Very low energy payback time Less sensitive to variations in semiconductor prices Greater potential for efficiency increase in the future compared to PV systems	Limited market – regions with high direct solar radiations, but not on rooftops Very difficult market entry – even for the lowest cost technologies Bankability and perception issues – due to shorter track record compared to PV Lack of technology standardization

Although research on cells, modules, and systems for CPV has been ongoing for decades, CPV only entered the market in the mid-2000s. With a total of more than 300 MWp, it is still a young and small player in the market for solar electricity generation compared to conventional flat-plate PV. This implies a lack of reliable data for market, prices, and technology standardization for industry.

However, the current trend in the CPV industry is more about the development of lower cost products with a much lower solar concentration to benefit of a quite low CapEx and which enables fast growth, instead of high concentration PV that requires a high initial investment. Taking advantage of being modular from the kW to the GW scale now, the concentrating photovoltaic sector is also less sensitive to variation in semiconductor prices, which implies interesting economic prospects.

Thus, even if this sector still lacks sufficient development inducing economic, technical or industrial gaps, it is essentially due to the novelty of this technology. But its development is constantly increasing, and with time, the difficulties of market entry regarding the competition with PV, the bankability and the perceptions issues are going to fade away, given the greater potential for efficiency increase of this technology compared to PV.

References

1. Shanks K., Sundaram S., Mallick T.K., Optics for concentrating photovoltaics: Trends, limits and opportunities for materials and design, Renewable and Sustainable Energy Reviews 60 (2016), 394–407.
2. Dreger M., Wiesenfarth M., Kisser A., Schmid T., Bett |A.W., Development and investigation of a CPV module with cassegrain mirror optics, AIP Conference Proceedings 1616 (2014), 177–182.

Chapter 5
Solar PVT Systems

5.1 Introduction

Solar energy is commonly converted using solar photovoltaic (PV) panels to produce electricity with efficiency about 15–20% and solar thermal (ST) panels to produce heat with efficiency up to 80%.

Under ideal conditions, a maximum of 20% of radiation is converted to electricity by current PV panel technologies while the unused infrared irradiation is converted into heat, which causes the electrical efficiency of solar cells to drop by about −0.4% per [°C] temperature rise above 25°C.

In order to increase the performance and efficiency of PV panels, the following actions have been considered:

- Keep the operating temperature at its ideal value otherwise the cells efficiency reduces.
- Recover the heat for use in appropriate applications.

A combination of these features can be reached using the so-called photovoltaic-thermal (PVT) panel, which combines PV and ST features to produce electricity and heat simultaneously. In fact, PVT integrates PV which converts ultraviolet and part of visible rays of the solar spectrum and ST which uses infrared rays of the spectrum to produce heat. A good deal of attention is being given to this technology due of its benefits as compared to solar thermal or PV system alone. Various theoretical, numerical and experimental studies have been made worldwide in the past few decades with the aim of maximizing the PVT panel technical and economical performances in terms of electrical as well as thermal outputs by considering:

- Various technologies of PV cells, heat exchangers…
- Optimum design of the layers composing the hybrid panels;

- Different fluids as an alternative to the conventional media that are water and air to harvest heat
- Materials such as PCM to store heat.

Any hybrid PVT system is composed of three main elements: the solar cells (PV laminate), a heat exchanger with one or multiple fluid channels, a heat extraction fluid. Other optional components can be added such as a concentrating system or a glazing layer.

There are a lot of parameters that can vary and characterize a PVT panel. These include:

- Type of PV laminate: crystalline silicon, CIGS, CdTe etc.
- Type of collector: flat plate liquid, flat plate air, concentrator, vacuum tube.
- Type of heat transfer medium: air, water, water-glycol mixture etc.
- Type of absorber: sheet and tube, a free-flow and a dual channel, roll bond.
- Type of insulation: covered, uncovered, with or without thermal insulation.
- Building integration.

PVT modules can thus be categorized in a number of ways: by their heat transfer medium (i.e. liquid or air), the relative positioning of the solar PV cells and the absorber, whether they are glazed or unglazed, whether they are insulated or not and how they are mounted (i.e. on-roof, in-roof, façade mounted, etc.)

5.2   Basic Principles

The commonly used panel collectors are generally separated in 2 distinct categories: PV for electricity and ST for water heating as depicted in the Fig. 5.1.

The combination of PV and ST technologies is considered an appealing concept which can be utilized to increase the performance of a solar system. In a PVT panel, the excess heat is extracted for water heating, air space heating, etc.

Fig. 5.1 Photovoltaic modules and a thermosiphon thermal collector

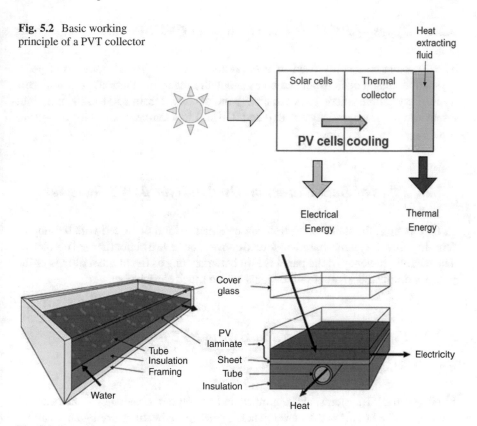

Fig. 5.2 Basic working principle of a PVT collector

Fig. 5.3 One-cover flat plate sheet and tube PVT collector

The combination of these two renewable energy technologies creates a unique opportunity for low carbon electricity and heat from the same source, the sun, making most efficient use of the surface area of the panels.

The basic principle of a PVT is summarized in Fig. 5.2.

A typical structure of a PVT is given Fig. 5.3.

This topology allows to cool actively the PV cells thanks to an effective thermal collector, transmitting the heat through a specific fluid (usually water or air) for further use.

The systems work typically between 40°C to 50°C, making them suitable for low temperature heating systems operating at the given temperature range. During summer, when the panels operate at peak conditions, temperatures can go up to 80°C which makes them suitable for hot water applications as well.

The flow distribution is one of the most influential factor on the electrical and thermal efficiency of the system. The mass flow rate is the main control variable to set the temperature of fluid to a desired value at the outlet of the PVT.

5.2.1 Types of PV Cells Used in Hybrid PVT Systems

The PV cells commonly used in PVT systems are the typical mono- and poly-crystalline silicon cells. Both can be used equally regarding thermal performances. Typical crystalline silicon cells can absorb incident light from 350 to 1200 nm (with a maximum sensitivity between 850 and 950 nm) in a temperature range of −40 to +85°C.

5.2.2 Types of Heat Extraction Used in Hybrid PVT Systems

In PVT panels, the fluid convection can be either natural or forced with a pump or fans. In general, systems based on forced convection have a better thermal efficiency. The overall efficiency of the panel is also better as long as the number of fans or the pump is optimized in order not to consume too much electrical power.

5.2.3 Types of Fluids Used in Hybrid PVT Modules

5.2.3.1 Air-Based Hybrid PVT Systems

In air-based PVT systems, air is pumped and heated in a channel or a cavity inside the panel. The hot air is then used generally for space heating, drying applications or dehumidification. In comparison with water-based PVT panels, air-based systems are less efficient because of the lower heat transfer coefficient, lower density, lower heat capacity and lower thermal conductivity of air. The thermal efficiency will not exceed 50% even for a well-design system. Air-based collectors have advantages though, such as no fluid freezing or leaking problems. Basic designs of air-based PVT are highlighted in Fig. 5.4.

In air-based PVT panels, air convection can be either natural or forced. In general, systems based on forced convection have a better thermal efficiency. The overall efficiency of the panel is also better as long as the number of fans is optimized in order to minimize electricity consumption.

5.2.3.2 BIPVT

BIPVT (Building Integrated PVT) refers to PVT systems that are integrated into the building envelope (Fig. 5.5). In addition to combining electrical and thermal power generation on the same surface, it can replace conventional roofing and thus compensate for the materials costs, to offer thermal insulation between the exterior and interior, and to provide a better thermal insulation on the sides of the system, in

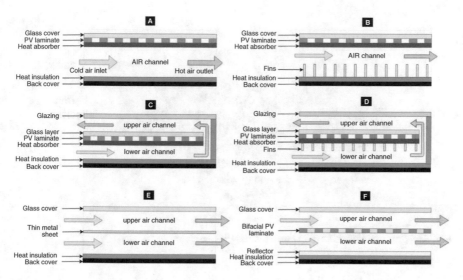

Fig. 5.4 Different topologies of air-based PVT collectors

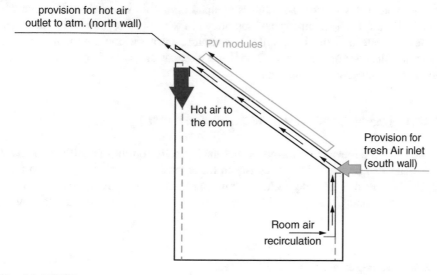

Fig. 5.5 BIPVT

comparison with PVT panels. At last, this solution may have a better aesthetic than panels fixed on a surface.

Some of the most promising applications of BIPVT systems include space heating, crop drying and electricity production.

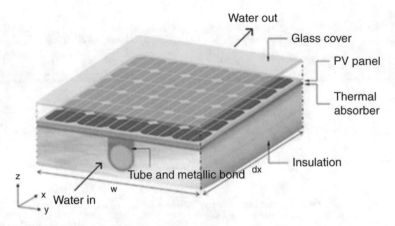

Fig. 5.6 Cross-section of a typical water-based PVT module

5.2.3.3   Water-Based PVT Systems

In water-based PVT systems, water is pumped through a water channel inside the panel (Fig. 5.6). The geometry and topology of the channel together with the water mass flow rate and the inlet water temperature have the most influence on the thermal efficiency of the system. The warm water at the outlet is mostly for space heating, cleaning, swimming pools.

5.2.3.4   Combined Water/Air (Bi-Fluid) Based PVT Systems

To enhance the thermal performance and the PV cells cooling capabilities, both air and water can be used simultaneously in the same PVT panel (Figs. 5.7 and 5.8). Using bi-fluid can provide higher output power as well as higher energy saving compared to using a single fluid which can be air, water or some phase change material.

5.2.3.5   Refrigerant-Based PVT Systems

Water-based PVT system cannot be used in geographical areas where the outdoor temperature falls under 0°C. In such context, the circulating fluid may be replaced by a refrigerant (e.g. glycol). In contrast with water-based panels, the heated fluid cannot be directly used and a heat exchanger is necessary. This kind of systems have a better thermal efficiency, but the enhanced performance may be cancelled by the losses inside the heat exchanger.

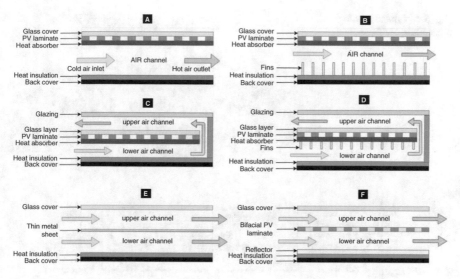

Fig. 5.7 Bi-fluid PVT systems

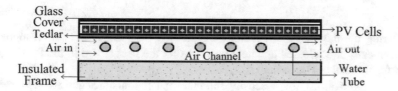

Fig. 5.8 Air-based PVT

5.2.4   Nanofluids-Based PVT Systems

In PVT systems, nanofluids are not only used as optical filter liquids, but also as a cooling fluid (Fig. 5.9). Nanofluids are composed of a base fluid in which nanoparticles (nano-meter or micro-meter scale) from metals or metal oxides are dispersed homogeneously. As base fluids, water, molten salt, ethylene-glycol, synthetic oils have been tested.

5.2.5   PCM (Phase Change Materials)-Based PVT Systems

Phase Change Materials (PCM) are capable of storing and releasing high amounts of latent heat when changing state (solid to liquid and vice-versa).
In order to improve the performance of PVT panels, the integration of a solid-liquid PCM static layer can be incorporated as depicted in Fig. 5.10.

Fig. 5.9 Typical topology of
a nanofluid PVT panel

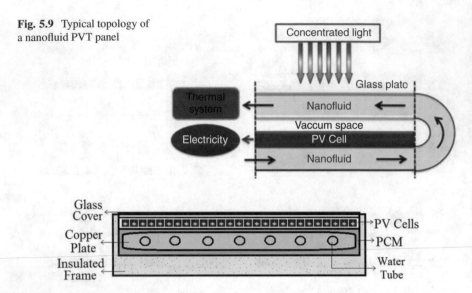

Fig. 5.10 Typical topology of a PCM-PVT panel

During the day, the PCM melts by absorbing heat from the sun. At night, the PCM freezes and releases its latent heat, so that the heat stored in the PCM can be used for thermal applications.

For PVT applications, low viscosity, high thermal conductance and adapted melting point are the key requirements for PCM.

5.2.6   Glazing

Glazing means addition of one or more layers and possibly one or more fluid channel(s) on top of the PV-cells. Adding glazing or not is an important option when designing a PVT panel since this defines its performance and its price.

In PVT panels, glazing acts like a protection, thus glazed collectors are more resilient to several environmental effects like soiling, breaking, moisture etc. Since the glazing layers absorb, reflect and refract a part of the solar irradiation, glazed panels prevents thermal losses and maximize the thermal output and the overall energy output. However, glazing reduces the electrical efficiency in an amount depending on the transmissivity of the glazing layer(s).

Regarding costs, PVT panels with glazing are more expensive than unglazed ones, because they use additional material. Unglazed PVTs are used for low temperature applications whereas glazed collectors are more suitable for hot water demand above 40°C.

5.3 Assessment

5.3.1 Summary of PVT Technologies

The following table summarizes the pros and cons of air-based versus water-based PVT considering different operational criteria.

Criteria	Air PVT	Water PVT
Assets	No fluid freezing or leaking	Heated water directly consumable
Limitations	Bad efficiency in cold climates	Freezing
		Leaking
		Overheating in hot climate
		Stagnation temperature
Suitable applications	Space heating, drying, dehumidification	Domestic hot water, preheating industrial processes
Electrical efficiency	9–12% (measured)	10–17% (measured)
Thermal efficiency	40–60% (measured)	50–70% (measured)
Most efficient design	Double-pass, with fins, multiple air entries, forced convection	Flat-plate, with pipes
Pumping, power needs	++	+++
Initial investment	+++	++++
Product warranty	NC	10 years

5.3.2 PVTs Versus PV+ST

In this section, we compare PVTs to separate use of ST + PV panels in a given space area.

Let us consider a surface S in m^2 with $x\%$ PV panels and $(1 - x)\%$ ST panels.

We denote C_{PV}, P_{PV} (resp. C_{ST}, P_{ST}) the cost and the power per m^2 of the PV (resp. ST) panels.

Similarly, we denote C_{PVT}, P_{PVT} the cost and the power per m^2 of the PVTs.

Then for a given surface S, the cumulative power (heat in Watt thermal W_{th} + electricity in Watt peak Wp) is:

- for PV + ST:

$$S[x P_{PV} + (1 - x)P_{ST}] \tag{5.1}$$

- for PVT:

$$SP_{PVT} \tag{5.2}$$

It is assumed that $P_{ST} > P_{PV}$ and $P_{ST} > P_{PVT}$ (which is generally the case). Therefore, PVTs provide more power if the following condition is satisfied:

$$x > \frac{P_{ST} - P_{PVT}}{P_{ST} - P_{PV}}. \tag{5.3}$$

For numerical illustration, we consider 2 typical PVTs available in the market with the following features:

- Dualsun: Unit area $= 1.66\,\text{m}^2$, $P_{PV} = P_{elec} = 280\,\text{Wp}$, $P_{ST} = 570\,\text{W}_{th}$. Therefore $P_{PVT} = (280 + 570)/1.66 = 511\,\text{W/m}^2$. In this case, the above condition can be written $x > 44.5\%$.
- Solimpeks: Unit area $= 1.37\,\text{m}^2$, $P_{PV} = P_{elec} = 190\,\text{Wp}$, $P_{ST} = 460\,\text{W}_{th}$. Therefore $P_{PVT} = (190 + 460)/1.37 = 474.5\,\text{W/m}^2$. The above condition writes $x > 50\%$.

In both cases, it can be seen that, for the same surface, if enough PVTs are installed, they can provide much more power than the combination of solar thermal and photovoltaic.

5.4   Modeling the PVT

Several studies have been conducted on the performance of a PVT collector both numerically and experimentally.
In this section, we will make use the following notations

S	area
I_{sun}	solar radiation
ℓ	thickness
M	mass
T	temperature
V_w	wind speed
L	length of the duct
h	heat transfer coefficient
W	width
d_{tub}	water inner tube diameter
$\dot{m}$	mass flow rate
α	absorptivity
β	packing factor
β_p	temperature coefficient
ε	emissivity

λ	thermal conductivity
μ	dynamic viscosity
ρ	density
η	efficiency
τ	transmissivity
am	ambient
cel	PV cell
gla	glass cover
ins	insulator
air	air fluid
wat	water fluid
abs	absorber
ted	tedlar
cv	convective
cd	conductive
rd	radiative

Also, let us recall that the electric power E_{el} produced by the PV part of the PVT is given by

$$E_{el} = I_{sun} S \eta_{cel} = I_{sun} S \eta_{ref}[1 - \beta_p(T_{cel} - T_{cel,ref})] \tag{5.4}$$

which will be used in the sequel.

5.4.1   Modeling the Air-Based PVT

We consider an air-based collector illustrated Fig. 5.11 composed of a glass cover, air channel, PV cells, and insulator

The governing dynamic energy balance equations for each layer (glass cover, PV cell layer, insulator) can respectively be written as [1]:

$$M_{gla} C_{gla} \frac{dT_{gla}}{dt} = S[\alpha_{gla} I_{sun} + h_{gla}^{rd}(T_{sky} - T_{gla})$$

$$+ h_{amb/gla}^{cv}(T_{amb} - T_{gla}) - h_{gla/air}^{cv}(T_{gla} - T_{air})$$

Fig. 5.11 Air-based configuration

$$+ h^{rd}_{gla,cel}(T_{cel} - T_{gla})] \tag{5.5}$$

$$M_{cel}C_{cel}\frac{dT_{cel}}{dt} = S[\tau_{gla}\alpha_{cel}I_{sun}\beta + h^{cv}_{cel,air}(T_{air} - T_{cel}) + h^{rd}_{gla/cel}(T_{gla} - T_{cel})$$

$$+ h^{cv}_{cel/ins}(T_{ins} - T_{cel})] - E_{el} \tag{5.6}$$

$$M_{ins}C_{ins}\frac{dT_{ins}}{dt} = S[h^{cd}_{cel/ins}(T_{cel} - T_{ins}) + h^{cv}_{ins/amb}(T_{amb} - T_{ins})] \tag{5.7}$$

For air channel, we have

$$\dot{m}_{air}C_{air}(T_{air,out} - T_{air,in}) = WL[h^{cv}_{gla/air}(T_{gla} - T_{air}) + h^{cv}_{cel/air}(T_{cel} - T_{air})]$$

where

$$T_{air} = \frac{T_{air,in} + T_{air,out}}{2} \tag{5.8}$$

$T_{air,out}$ (resp. $T_{air,in}$) is the inlet (resp. outlet) temperature in the air channel.

5.4.2   Modeling the Water-Based PVT

In this section, we give the dynamic thermal model of a PVT using the energy balance principle for each component and layer of a PVT. We consider a water-based collector made of different layers as presented in Fig. 5.12: a glass cover, PV cells, tedlar, absorber, metallic tubes through which water flows and an insulator.

The dynamic energy balance takes into account the masses and the heat capacities of each layer forming the PVT collector to derive the temperatures of each component [2]:

$$M_{gla}C_{gla}\frac{dT_{gla}}{dt} = S[\alpha_{gla}I_{sun} + h^{rd}_{gla/sky}(T_{sky} - T_{gla})$$

$$+ h^{cd}_{gla/cel}(T_{cel} - T_{gla}) + h^{cv}_{amb/gla}(T_{amb} - T_{gla}] \tag{5.9}$$

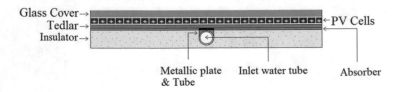

Fig. 5.12 Water-based configuration

$$M_{cel}C_{cel}\frac{dT_{cel}}{dt} = S[\tau_{gla}\alpha_{cel}\beta I_{sun}$$

$$+ h_{gla/cel}^{cd}(T_{gla} - T_{cel}) + h_{cel/ted}^{cd}(T_{ted} - T_{cel})] - E_{el} \tag{5.10}$$

$$M_{ted}C_{ted}\frac{dT_{ted}}{dt} = S[\tau_{gla}\alpha_{ted}(1-\beta)I_{sun} + h_{cel/ted}^{cd}(T_{cel} - T_{ted})]$$

$$+ S_{ted/abs}h_{ted/abs}^{cd}(T_{abs} - T_{ted}) \tag{5.11}$$

$$M_{abs}C_{abs}\frac{dT_{abs}}{dt} = S_{ted/abs}h_{ted/abs}^{cd}(T_{ted} - T_{abs}) + S_{abs/tub}h_{abs/tub}^{cd}(T_{tub} - T_{abs})$$

$$+ S_{abs/ins}h_{abs/ins}^{cd}(T_{ins} - T_{abs}) \tag{5.12}$$

$$M_{tub}C_{tub}\frac{dT_{tub}}{dt} = S_{abs/tub}h_{abs/tub}^{cd}(T_{abs} - T_{tub}) + S_{tub/ins}h_{tub/ins}^{cd}(T_{ins} - T_{tub})$$

$$+ S_{tub/wat}h_{tub/wat}^{cv}(T_{wat} - T_{tub}) \tag{5.13}$$

$$M_{wat}C_{wat}\frac{dT_{wat}}{dt} = S_{tub/wat}h_{tub/wat}^{cv}(T_{tub} - T_{wat})$$

$$+ \dot{m}_{wat}C_{wat}(T_{wat,in} - T_{wat,out}) \tag{5.14}$$

$$M_{ins}C_{ins}\frac{dT_{ins}}{dt} = S_{tub/ins}h_{tub/ins}^{cd}(T_{tub} - T_{ins}) + S_{ins/abs}h_{ins/abs}^{cd}(T_{abs} - T_{ins})$$

$$+ Sh_{ins/amb}^{cv}(T_{amb} - T_{ins}) \tag{5.15}$$

where E_{el} is the fraction of absorbed solar radiation that is converted into electricity given in (5.4) and where (T_{wat}) is the arithmetic mean of the inlet $(T_{wat,in})$ and outlet $(T_{wat,out})$ temperatures which is given by:

$$T_{wat} = \frac{T_{wat,in} + T_{wat,out}}{2} \tag{5.16}$$

5.4.3 Heat Transfer Coefficients

5.4.3.1 Conduction

The conductive heat transfer coefficient between two layers (i, j) is given as:

$$h_{i/j}^{cd} = \frac{1}{\dfrac{\ell_i}{\lambda_i} + \dfrac{\ell_j}{\lambda_j}} \tag{5.17}$$

5.4.3.2   Convection

In an air-based type, the convective heat transfer coefficients between air–tedlar and air–insulator can be calculated by the following relation:

$$h_{air}^{cv} = N_u \frac{\lambda_{air}}{d} \tag{5.18}$$

where the Nusselt number N_u is;

$$N_u = 0.023 R_e^{0.8} P^{0.4} \tag{5.19}$$

and Reynold number R_e and Prandtl number P are given as;

$$R_e = \frac{\rho_{air} d}{\mu_{air}} V_{air} \tag{5.20}$$

and

$$P_{abs} = \frac{\mu_{air}}{\lambda_{air}} C_{air} \tag{5.21}$$

The metallic tubes in water-based types are pressed within the absorber and are attached through a bond to the absorber plate. The heat transfer coefficient adds inversely in series. The heat transfer coefficient in the pipe h_{wat} is assumed laminar, that is $Re < 2300$ and

$$h_{wat} = 4.364 \frac{\lambda_{wat}}{d_{tub}}. \tag{5.22}$$

Generally, the sky temperature T_{sky} is calculated using the Swinbank equation which associates local air temperature with the sky temperature and requires no knowledge of the dew point temperature.

$$T_{sky} = 0.05 T_{amb}^{1.5}. \tag{5.23}$$

For an overcast day, sky temperature is equivalent to ambient temperature.
Convective heat coefficient highly depends on the wind speed. For a vertical wall with natural air flow and with temperatures close to ambient temperature (300 K), the thermal convection coefficient for air is around 10 W/(m^2 K). An approximation formula gives a valid value between 5 and 18 m/s for air circulating around the object:

$$h_{amb}^{cv} = 10.45 - V_w + 10\sqrt{V_w}$$

Alternative correlations can be found in the literature depending on the location, for instance :

$$h_{amb}^{cv} = h_1 + h_2 V_w, \quad h_1 = 6, \ h_2 = 3. \tag{5.24}$$

5.4.3.3 Radiation

Radiative heat transfer coefficient between glass and sky is:

$$h_{gla/sky}^{rd} = \sigma \varepsilon_{gla} \frac{T_{gla}^4 - T_{sky}^4}{T_{gla} - T_{amb}} \tag{5.25}$$

The heat transfer coefficient due to radiation between two infinite parallel plates (i, j) is given by:

$$h_{i/j}^{rd} = \sigma \left[\frac{(T_i + T_j)(T_i^2 + T_j^2)}{\dfrac{1}{\varepsilon_i} + \dfrac{1}{\varepsilon_j} - 1} \right] \tag{5.26}$$

Remark: An extra glass cover with an enclosed air space can be added over the PV module (glazing). The glazed PVT model will have two extra equations i.e. one for extra glass and another for enclosed air space. The heat transfer occurs by convection in the closed air space whereas heat transfer occurs by radiation between the top and bottom glass layers. This extra glass cover increases the temperature of all the layers of the PVT collector which will result in low PV efficiency.

5.4.4 State Space Description

The thermal model can be expressed using a general nonlinear state space form

$$\begin{cases} \dot{X} = f(X, U, V) \\ Y = g(X, U) \end{cases}, \tag{5.27}$$

where the state X contains the temperatures of each layer as state variables. For instance, in the case of water-based PVT, we have

$$X = \begin{bmatrix} T_{gla} & T_{cel} & T_{ted} & T_{abs} & T_{tub} & T_{wat} & T_{ins} \end{bmatrix}^T$$

whereas the control input U is the mass flow rate $\dot{m}_{air}$ and V are the perturbations (uncontrolled inputs),

$$V = \begin{bmatrix} T_{amb} & I_{sun} & T_{wat,in} \end{bmatrix}^T$$

and the output Y is the outlet temperature ($T_{wat,out}$).

This nonlinear form is due to the heat transfer coefficient which depend on the states. The state space description is useful for simulation of the system dynamic behavior.

If we consider an operating point defined by the subscript 'o', and if we denote by (X_o and U_o) the equilibrium point for the state and the input, then for small variations $x = X - X_o$, $u = U - U_o$, $v = V - V_o$ and $y = Y - Y_o$, we can use a linearized state-space representation described as :

$$\begin{cases} \dot{x} = Ax + Bu + Gv \\ y = Cx + Du \end{cases} , \tag{5.28}$$

where

$$A = \left.\frac{\partial f}{\partial X}\right|_{X_o,U_o,V_o}, \qquad B = \left.\frac{\partial f}{\partial U}\right|_{X_o,U_o,V_o}, \qquad G = \left.\frac{\partial f}{\partial V}\right|_{X_o,U_o,V_o} \tag{5.29}$$

$$C = \left.\frac{\partial g}{\partial X}\right|_{X_o,U_o}, \qquad D = \left.\frac{\partial g}{\partial U}\right|_{X_o,U_o} \tag{5.30}$$

This linear form is useful for control using well known linear state feedback techniques such as pole placement, LQR,... [2].

5.4.5 Simulation

For simulation, we use typical parameters provided in Table 5.1. Regarding the areas, examples for a typical PVT are given for each layer:

- glass: $2\,m^2$;
- cell: $1.6\,m^2$;
- insulator: $2\,m^2$.

With these values, for a standard summer day as in Fig. 5.13, the temperatures evolution of a water-based PVT collector are presented in Fig. 5.14.

One can also easily check that the outlet temperature of water decreases when the water flow increases (Fig. 5.15) or when wind speed increases (Fig. 5.16).

Table 5.1 Example of geometric and thermal values for PVT

Parameters	Value
α_{gla}	0.06
L_{gla}	0.004 m
C_{gla}	670 J/(kg K)
λ_{gla}	1.1 W/(m K)
α_{cel}	0.84
L_{cel}	0.0002 m
C_{cel}	900 J/(kg K)
λ_{cel}	140 W/(m K)
C_{ins}	670 J/(kg K)
λ_{ins}	0.034 W/(m K)
β	0.88
β_p	0.0045
ε_{gla}	0.88
ε_{cel}	0.96
τ_{gla}	0.93

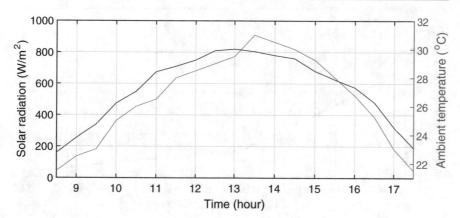

Fig. 5.13 Solar radiation and ambient temperature versus time

5.5   Examples of Hybrid PVT Products

The solar-thermal technology market is expanding rapidly due to its benefits but at present, the market is comparatively small. In the future, these PVT systems coupled to heat pumps can cover more than 60% of heating and 50% cooling demand for urban residential buildings.

Examples of PVT modules available in the market are given in Table 5.2 below with their specifications such as working medium, weight, solar cell type, covered/uncovered, area, electric and thermal yield per panel.

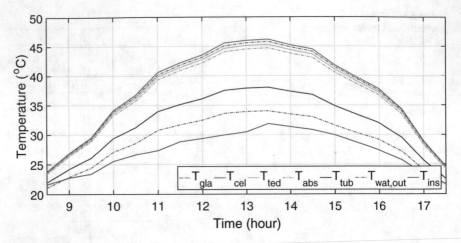

Fig. 5.14 Water outlet temperature versus time

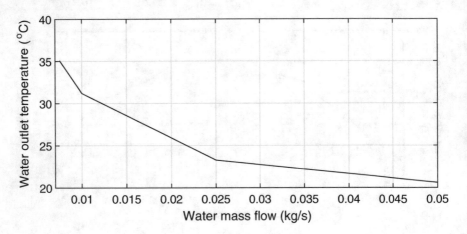

Fig. 5.15 Temperature of water at the outlet versus water flow

5.6   Applications

PVT panels combine two well established renewable energy technologies into one integrated component that generates both low carbon electricity and heat from the same renewable energy source. The advantages of the technology can be summarized as;

- Higher energy yield per m^2, allowing a higher percentage of energy to be self-generated on properties with smaller roofs
- Provide a more uniform look to the building than separate solar PV and solar thermal systems

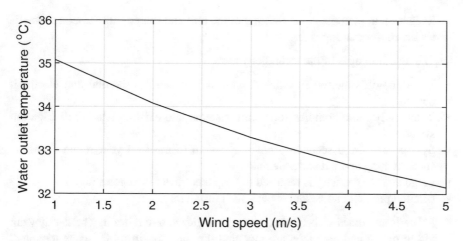

Fig. 5.16 Water outlet temperature versus wind speed

Table 5.2 Examples of PVT products available in the market and their specifications

Name of product and specifications	Example
Solar one hybrid collector (3F Solar) • Uncovered flat plate with thermal insulation • Water-based • Solar cells: mono cSi • Collector Area: $1.69\,m^2$ • Weight: 42 kg • Electricity yield: 265 Wp • Thermal yield: $825\,W_{th}$	
Twin Solar (SYSTOVI) • Air-based • Solar cells: mono cSi • Collector Area: $1.5\,m^2$ • Weight: 20.7 kg • Ventilated PV module • Heat recovery system • Heat during the winter, cool air over the summer • Temperature controlled • Electrical yield: 250 Wp • Thermal yield: $450\,W_{th}$	

- Can generate a lot of low grade heat from a domestic scale system and complements a number of other low carbon heating technologies
- Has the potential to make a significant contribution to heating and hot water when designed and installed as part of an energy efficient new build

- Reduction in material costs within module unit and shared infrastructure materials i.e. roof mounting

The disadvantages of the technology are:

- The trade-off required in efficiency terms between the thermal and electrical output
- Technically more complex to retrofit systems into existing heating distribution systems
- Systems require a secondary heating system to meet all year round domestic space heating and hot water demand.
- Large storage capacity is required to keep electrical efficiencies high
- Limited product choice and experienced installers

PV cells and the absorber, whether they are glazed or unglazed, whether they are insulated or not and how they are mounted (i.e. on-roof, in-roof, façade mounted, etc.).

The possible growth markets for PVT in the longer term have been identified as installations that have the potential combination of:

- Where there is limited roof space for installation (including zero carbon homes).
- High hot water demand (particularly in the summer).

The establishment of standards to assess the performance of PVT and provide a framework on how to define product characteristics is key to allow products to be impartially compared and utilized.

PVT system has a number of applications such as solar greenhouse, trigeneration, air conditioning, desalination, district heating/cooling, drying, heat storage, refrigeration, and industrial applications [3–9]. The applications can be divided according to their temperatures as mentioned in the IEA SHC programme report.

- High temperature applications that require a temperature above 80 °C are used for some industrial processes e.g desalination and agro-industrial processes.
- Medium-temperature applications are used in buildings for domestic hot water and space heating where the temperature up to 80 °C is required. Moreover, PVT integrated with a heat pump can be used for desiccant heating where 50–60 °C temperature is required.
- Low-temperature applications take in heat pump systems and heating swimming pools or spas up to 50 °C.

To illustrate the use of PVT for a private home (Fig. 5.17), let us consider the following assumptions:

- n PVT panels of 250 Wp electricity and 500 W_{th} thermal power.
- Water tank with storage capacity of 200 L.
- 5 h of daily peak sunshine.

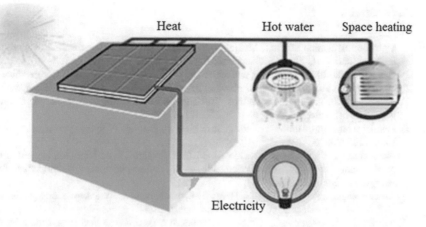

Fig. 5.17 PVT roof-top array supplying heating and electricity to building

The energy E (J) needed to increase by ΔT (°K) the temperature of m (kg) of water can be expressed as

$$E = mC\Delta T$$

Taking into account the fact that $1\,\text{Wh} = 3600\,\text{J}$ and that for water, $C = 4185\,\text{J/(kg} \times \text{K)}$, we get E in Wh as:

$$E = 1.16V\Delta T$$

where V is the volume of water in (L)

With the above given values, the daily thermal energy delivered by the n panels is $5 \times 500 \times n = 2500n$ Wh. The increase of temperature of the water tank is approximated as

$$\Delta T = nE/(1.16V) = 2500n/(1.16 \times 200) = 10.76n.$$

Obviously, these calculations assume no losses and enough time to reach steady state. However, they can be useful as a quick way to evaluate the temperature increase provided by the thermal energy of the PVTs.

Besides, the PV panels part provides a daily electric energy of $250 \times n \times 5 = 1250n$ Wp. This energy can also be used to heat the water in the tank and thus be converted as heat storage to complement the thermal part of the PVT when necessary.

Acknowledgments Ahmed RACHID is thankful to Dr. Zain Ul Abdin for his contribution to this chapter.

References

1. Mohamed El Amine Slimani, Madjid Amirat, Ildikó Kurucz, Sofiane Bahria, Abderrahmane Hamidat, and Wafa Braham Chaouch. A detailed thermal-electrical model of three photovoltaic/thermal (pv/t) hybrid air collectors and photovoltaic (pv) module: Comparative study under algiers climatic conditions. *Energy conversion and management*, 133:458–476, 2017.
2. Zain Ul Abdin and Ahmed Rachid. Bond graph modeling of a water-based photovoltaic thermal (pv/t) collector. *Solar Energy*, 220:571–577, 2021.
3. Zain Ul Abdin and Ahmed Rachid. A survey on applications of hybrid pv/t panels. *Energies*, 14(4):1205, 2021.
4. María Herrando, Christos N Markides, and Klaus Hellgardt. A uk-based assessment of hybrid pv and solar-thermal systems for domestic heating and power: system performance. *Applied Energy*, 122:288–309, 2014.
5. Aleksis Baggenstos Daniel Zenhäusern, Evelyn Bamberger. Wrap-up. energy systems with photovoltaic thermal solar collectors. *Final report*, 2017.
6. de Jong M. de Keizer A.C., Bottse J. Pvt benchmark: An overview of pvt modules on the european market and the barriers and opportunities for the dutch market. 2018.
7. Lars Lisell Jay Burch Dennis Jones David Heinicke Jesse Dean, Peter McNutt. Photovoltaic-thermal new technology demonstration. 2015.
8. R. Saidur A. S. Abdelrazik, F. A. Al-Sulaiman and R. Ben-Mansour. A review on recent development for the design and packaging of hybrid photovoltaic/thermal (pv/t) solar systems. *Renew. Sustain. Energy Rev.*, 95:110–129, 2018.
9. Evidence gathering – low carbon heating technologies – hybrid solar photovoltaic thermal panels. department for business, energy and industrial strategy. 2016.

Chapter 6
Smart Grids and Solar Energy

6.1 Introduction to Smart Grids

Traditional power grids, which are still prevalent today, were constructed based on the predominance of bulk centralised generation. Their design involved the assumption of unidirectional power flows through high-voltage power lines and lower voltage distribution networks.

Because of the need to reduce environmental pollution from power generation, which contributes negatively to global warming and climate change, and the gradual depletion of fossil fuel resources, a trend has emerged in the last few decades of generating power locally at distribution voltage level. This power generation may involve conventional generation, such as small diesel generators, and renewable energy sources, such as solar PV, wind turbines or small hydro. These sources inject power into the utility distribution network. This type of power generation is known as *distributed generation*. These installations are not centrally planned but instead are funded and deployed by local entities, including the residential, commercial, industry and local government sectors. Their connection is often subject to permission by the utility company. However, as will be further discussed later in this chapter, the penetration of distributed generation increases the complexity of power grids and presents significant stability, control and protection challenges. Additionally, the traditional power grid faces a number of other challenges [1–3], which include:

- *Ageing infrastructures:* large portions of today's infrastructure date from the 1960s or even earlier and are nearing the end of their useful lives. Furthermore, during peak demand, equipment is subjected to high stress.
- *Integrating intermittent energy sources:* the integration at large scale of intermittent energy sources to the existing grid is likely to strain it. As a result, the intermittent nature of these sources must be counterbalanced by increased grid

A. Rachid et al., *Solar Energy Engineering and Applications*, Power Systems,
https://doi.org/10.1007/978-3-031-20830-0_6

intelligence, baseload power supply (such as nuclear power plants), and energy storage.

- *Security of supply and increase in energy needs:* electricity transmission and distribution must be efficient and reliable in order for modern societies and economies to function well. Moreover, electricity demand is steadily increasing in many countries.
- *Sustainability:* there is pressure on and commitment from governments to reduce CO_2 emissions through the adoption of alternative energy sources. Governments are also putting in place regulations to increase energy efficiency.
- *The need for lower energy prices:* To bring down energy prices, regulators are pushing for more competition. As a result, utilities must spend in information and communication technologies in order to remain profitable and still be able to invest in infrastructure.
- *Challenges faced by utility companies*: these challenges include high power system loading, increasing distance between generation and consumption, additional and new consumption models (electric vehicles, smart buildings), central power generation in parallel with distributed generation, increasing costs and regulatory pressures, utility unbundling, increased energy trading, and the need for transparent consumption and pricing for the consumer.

The above challenges act as drivers for modernisation through the development and deployment of what is known as smart grid technologies. However, the need for modernisation of the power grid varies between countries and region, depending on the state of the national or regional transmission and distribution grids, the level of penetration of variable renewable energy sources, and the trends in electricity demand. Although there is no universally accepted definition of smart grid, the following definition is suitable for this book:

> A smart grid is an electricity network that uses comprehensive computing, sensing, power electronics, control and communication technologies to intelligently integrate the actions of all users connected to it to efficiently deliver sustainable, economic and secure electricity supplies.

Smart grid technologies have the potential to address the challenges faced by the traditional grid. Various smart grid innovations, such as smart meter technology, are already being deployed and are contributing to the modernization of electric power grids. A smart meter, an important element of smart grid, is a new kind of energy meter that can send readings to the utility company via a wireless or wired communications infrastructure. However, as power systems are so large, it is impossible to upgrade them with smart grid technology in a short period of time. Therefore, the changes associated with smart grid deployments tend to occur slowly. An example of this slow pace is the deployment of smart meters, which is taking years in most countries.

As illustrated in Fig. 6.1, the traditional grid is characterised by centralised, bulk generation, heavy reliance on coal and oil, limited automation, limited situational awareness, and the lack of data by consumers to manage their energy usage. In traditional grids, network information is typically collected from a limited number

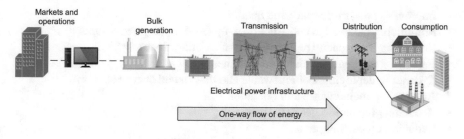

Fig. 6.1 Illustration of the traditional grid

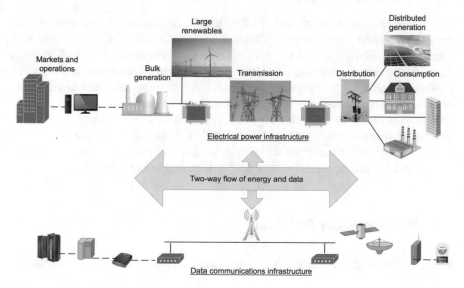

Fig. 6.2 Illustration of the smart grid

of network nodes or substations. Communication has been part of the power grid for a long time, but only for specific applications or purposes [1].

In contrast, referring to Fig. 6.2, smart grid is characterised by a modern and comprehensive communications infrastructure that operates alongside the power infrastructure. A further aspect that is common in smart grids is the growing connection of variable renewable energy. The above factors result in a two-way flow of electricity and information. Smart grids use digital and communications technologies to accommodate and manage the bi-directional flow of data between end-users and utility companies, between elements of infrastructure and the utility companies, and between different organisations involved in the grid, as well as the bi-directional flow of power that may occur at different branches of the network.

The deployment of smart grid technologies brings several benefits to the power grids themselves, to key players in the electric energy system, environment, economy, and society. These benefits include [1–3]:

- Improving power reliability and quality
- Minimizing the need to construct back-up (peak load) power plants
- Enhancing the capacity and efficiency of existing electric grid
- Improving resilience to disruption and being self-healing
- Expanding deployment of renewable and distributed energy sources
- Automating maintenance and operation
- Reducing greenhouse gas emissions
- Reducing oil consumption
- Enabling transition to plug-in electric vehicles (EVs)
- Increasing consumer choice

There are aspects of smart grid that facilitate the integration of solar photovoltaic plants or may affect the operation of such plants in different ways. This chapter will focus on critical aspects of the integration of solar photovoltaic plants of different sizes into the power grid. As a vital element in such interconnection is the inverter (a power electronics device introduced in Chap. 2 that converts DC electricity into AC electricity), we introduce the inverter architectures typically used to integrate solar photovoltaic plants into the grid. We will also look at the criteria and approaches for grid integration and the role of energy storage in facilitating the integration of solar photovoltaic plants. Finally, we will also discuss ways of selling solar photovoltaic energy and the available mechanisms for participating in energy markets.

6.2   Types of Solar PV Plants

For the purposes of this chapter, it is a good idea to classify solar photovoltaic plants as follows.

- *Utility-scale solar farms.* These plants occupy large areas and may have a capacity between a few to hundreds of MW. The power produced by these plants can be injected into the electric grid at transmission or distribution voltage levels. These plants are not dispatchable, but the power system operator may curtail their production in some cases. Figure 6.3 shows an aerial view of a utility-scale solar farm.
- *Distributed PV generation.* Small residential and commercial plants range between a few to hundreds of kW. They are located on rooftops or land at homes, commercial, municipal or industrial property. These resources are interfaced to the grid at distribution voltage levels. They may be installed *behind the meter*, so that the electricity is generated for self-consumption, with occasional export of power to the grid. Other plants may be installed *in front of the meter* so that the power is generated to export it to the distribution grid, rather than for self-consumption. Figure 6.4 shows a solar array on a large industrial rooftop as an illustration of a distributed PV generation facility.

Fig. 6.3 Aerial view of a utility-scale solar farm

Fig. 6.4 View of a large industrial rooftop with a solar photovoltaic array as an example of a distributed generation facility

6.3   Inverter Architectures

The integration of PV systems to the electricity grid requires grid-connected inverters. This type of inverter takes the DC input from the PV array and produces the AC output required by the utility grid. A grid-connected inverter will only function when the grid is operating within particular voltage and frequency ranges. Many, but not all, grid-connected inverters are designed to allow exporting AC power to the grid.

In addition to converting the DC electricity from the PV array into AC electricity, grid-connected inverters ensure that the AC current injected into the grid is at the right frequency and only has an acceptable harmonic distortion. They also employ maximum power point tracking (MPPT) technology to extract the greatest possible amount of power from the PV array given the sunlight conditions. Moreover, a grid-connected inverter has appropriate protections that disconnect it from the grid when the operating conditions are outside its tolerances unless it is equipped with fault-ride through capabilities to meet local grid standards.

Figure 6.5 shows a set of three grid-connected central solar inverters at Westhampnett solar farm in West Sussex, England. These inverters interface the 7.4 MWp solar photovoltaic array with the three-phase electricity grid.

Fig. 6.5 Central solar inverters at Westhampnett solar farm in West Sussex, England. These inverters interface the 7.4 MWp solar photovoltaic array with the three-phase electricity grid

Different categories of grid-connected inverters are available, depending on the architecture of the solar array and the way it is connected to the inverter. These categories, which include string, multi-string, central and modular inverters, are discussed below [4]. Some inverters provide a single-phase AC output, while others provide a three-phase AC output (but not modular inverters). It is also possible to use three single-phase Y-connected inverters (one for each phase) to interface with the three-phase electricity network.

6.3.1   String Inverters

A string inverter (Fig. 6.6) has a single MPPT input and is connected to one or more strings of solar modules. A typical string of solar modules consists of around 10-15 modules connected in series. The number of solar modules in series that can be connected to a string inverter depends upon the input voltage rating of the inverter. Two or more strings in parallel can be connected to the MPPT input of the string inverter.

This type of solar inverter has been prevalent in the solar photovoltaic industry. It is the most tried-and-tested and cost-efficient inverter available in the market. It only optimises the power output at a string level, not at an individual module level. Typically, the power rating of string inverters is up to 10 kW.

String inverters are very popular because they are smaller than central inverters. They monitor the solar installation at a string level. However, they could be a single point of failure, given that when the inverter breaks down, one or more strings of modules become disconnected.

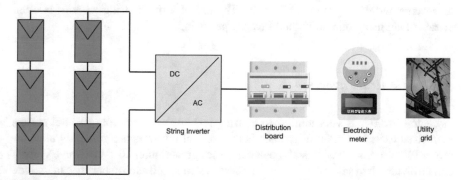

Fig. 6.6 Illustration of the string inverter architecture, showing two strings in parallel connected to the MPPT input of the inverter

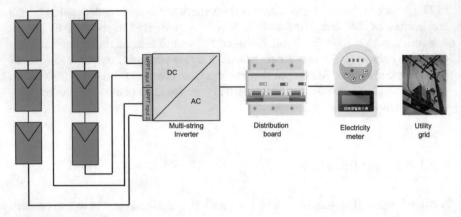

Fig. 6.7 Illustration of the multi-string inverter architecture, showing two strings connected independently to each of the MPPT inputs of the inverter

6.3.2   Multi-String Inverters

A multi-string inverter (Fig. 6.7) has more than one MPPT inputs. One string of solar modules can be connected to each MTTP input, such that the PV array is divided into multiple strings. These inverters have the advantage that if there are modules that are oriented in different directions (or are subject to different shading conditions), then the array can be divided into strings so that modules that belong to the same string are all subject to similar shading (or are oriented in the same direction). Having each string connected to a dedicated MPPT input allows the energy yield from the system to be greater than if the strings were connected to an inverter with only one MPPT input. The use of a multi-string inverter is often cheaper than using one individual inverter per string.

6.3.3   Central Inverters

A central inverter is very similar to a string inverter with multiple parallel strings connected to it. The main difference is that central inverters are generally used for larger PV systems, usually with peak capacities greater than 10 kWp. The PV array can be divided into several subarrays in these systems, and each subarray can include several strings. In some systems, just one large inverter is adequate for the whole PV array. Alternatively, the single central inverter may contain several smaller multi-string inverters combined to produce a single AC electrical output. In some cases, a central inverter can consist of several smaller inverters that can automatically be selected to operate depending on the amount of power that can be generated from the available sunlight. This configuration often improves the efficiency of the central inverter unit. However, a disadvantage of central inverters is that there is no redundancy, such that when the inverter fails, the whole PV array goes out of service.

6.3.4 Modular Inverters

Modular inverters (Fig. 6.8), or micro-inverters, are small inverters with a power rating usually in the order of the hundreds of watts that are used to convert from DC to AC the electricity produced by a single solar PV module. They are designed to be mounted at the back of the PV module. The key advantages of the modular inverter are that they remove the requirement for DC cabling from the array as each module has an AC output. The AC cables coming from the modular inverters can be connected in parallel and then connected to the grid at the appropriate location. Modular inverters are small, light, and they allow extra solar modules/inverter sets to be added to the system at minimum cost. A disadvantage of modular inverters is that if the inverter fails, repairing or replacing it involves reaching the location of the corresponding PV module and removing the module from the array to access the inverter behind it. Moreover, as the temperature behind solar modules can become rather high in hot climates, the operation of modular inverters can be negatively affected.

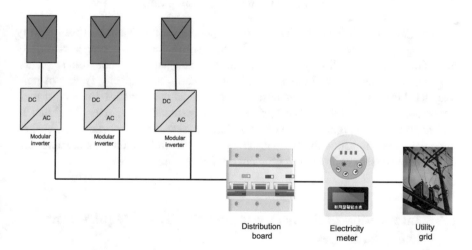

Fig. 6.8 Illustration of the modular inverter architecture, showing three PV modules each with their own modular inverter

6.4   Supply and Demand Balancing

In order to balance supply and demand, power system operators need to decide which generation units are going operate in a given day. This decision is known as the *unit commitment problem*, which is typically solved once a day. The operators also need to decide how much active power will be produced by each generation unit, considering economic operation (e.g. minimising total generation costs). This is known as the *economic dispatch problem*, which is typically solved once every 30 min. Finally, the power system needs to employ *automatic generation control* (AGC), which performs more frequent adjustments to generation levels.

The dispatchability of an electricity generation source refers to the source's ability to be controlled in response to system requirements, such as variation in demand at the request of the power grid operator. Conventional generators, such as gas, coal and nuclear plants, are *dispatchable*, as are some renewable energy plants (e.g. concentrated solar power with storage, geothermal power and hydro), while other renewable energy plants are *non-dispatchable* (e.g. solar photovoltaic and wind power facilities).

The capacity factor is the ratio of actual electricity produced by a power plant in a year to the electricity that the plant could theoretically produce if operated continuously at full power during the same period:

$$\text{Capacity Factor} = \frac{\text{Actual energy produced in 1 year}}{\text{Maximum theoretical energy produced in 1 year}} \qquad (6.1)$$

Apart from maintenance shutdowns, dispatchable renewables are constantly available for production and offer high capacity factors. In contrast, electricity generation from non-dispatchable renewables, such as solar photovoltaic plants, depends on meteorological conditions. Consequently, the capacity factors of non-dispatchable renewables are modest, and grid operators cannot fully plan their generation. As a result, an appropriate backup generation capacity is needed in power grids with a significant share of variable renewables. Unfortunately, backup generators usually run with low loads compared with their rated capacity, and thus they do not operate at their most efficient operating point.

6.5   Integration of Solar Power Into the Grid: Requirements and Challenges

The share of variable renewables in overall power generation is rapidly increasing in many countries. Figures 6.9 and 6.10 show, respectively, the evolution of the share of electricity production from solar and wind energy in various countries between 2000 and 2021.

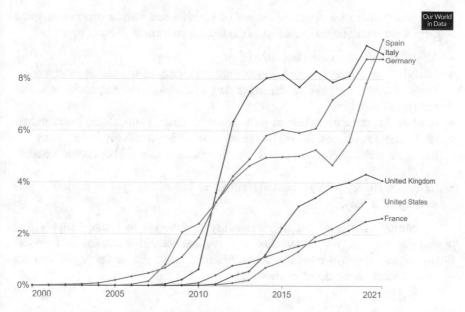

Fig. 6.9 Share of electricity production from solar energy in various countries between 2000 and 2021. Source: Our world in data based on BP statistical review of world energy & ember (2022). OurWorldData.org/energy · CC BY

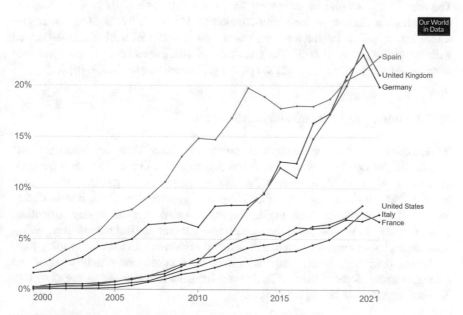

Fig. 6.10 Share of electricity production from wind energy in various countries between 2000 and 2021. Source: Our world in data based on BP statistical review of world energy & ember (2022). OurWorldData.org/energy · CC BY

The integration of a significant share of variable renewables into power grids requires a substantial transformation of the existing networks in order to:

- Allow for a bi-directional flow of energy;
- Establish an efficient electricity-demand and grid management mechanisms aimed at reducing peak loads, improving grid flexibility, responsiveness and security of supply
- Improve the interconnection of grids at the regional, national and international level, aimed at increasing grid balancing capabilities, reliability and stability
- Introduce technologies and procedures to ensure proper grid operation stability and control
- Introduce energy storage capacity aimed at increasing system flexibility and security of supply.

By adding aspects of smart functionality to the grid to balance supply and demand and employing information and communication technologies to boost flexibility and improve reliability and efficiency, smart grid technologies can act as an enabler for these developments.

6.5.1   Challenges with the Integration of Solar Power

The deployment of photovoltaic solar farms is occurring rapidly in some regions, often bringing challenges for utility companies. However, by adopting smart grid technologies, utilities reduce the technical and operational difficulties associated with integrating solar energy. The most critical challenges utility companies face when integrating solar farms into the power grid are discussed below [5].

6.5.1.1   Balancing Demand and Generation

In regions where solar generation becomes significant, there are various issues that the power networks were not designed to address. One of the main issues is intermittency. As an illustration of the intermittence of solar generation, consider Fig. 6.11, which shows the total solar PV power generated in Great Britain during the month of July 2021. Note the day/night cycles and the day-to-day variability. Intermittence can be addressed through reserve generation, fast responding generators, dispatchable energy storage, and demand response schemes. To employ most of these approaches, utilities require sufficiently accurate and timely weather and solar generation forecasts and effective mechanisms to use those forecasts to take the most appropriate balancing actions.

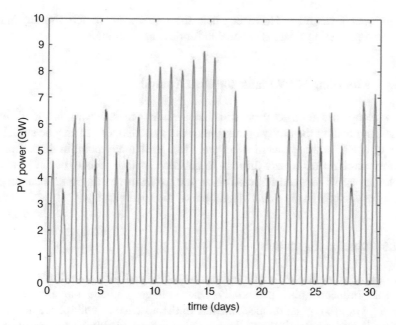

Fig. 6.11 Total grid-connected solar PV power generated in Great Britain during the month of July, 2021. Data points consist of half hourly averages. Note that the installed grid-connected PV capacity is estimated at 13.08 GWp. To have an idea on the relative contribution of solar, consider that the total power demand was, on average, 30.8 GW in 2021. Data source: Live PV website [6]

6.5.1.2   PV Hosting Capacity of Distribution Feeders

PV hosting capacity is the total PV power that can be injected on a distribution feeder without negative impacts on voltage, protection, power quality, and feeder reinforcement. The determination of the PV hosting capacity of distribution feeders is key to establishing their capability to support the integration of new solar farms, Furthermore, this depends on the technology used by the solar farms, their peak power capacity, location, and feeder characteristics. Estimating the hosting capacity of an individual feeder requires a detailed study that is not always possible or economical to perform. Then it is common to carry out such studies on a few selected feeders and translate the results into other feeders under the assumption that they perform similarly, which may not be entirely accurate.

6.5.1.3   Coordination of Protection Systems

The difficulty of protection coordination increases considerably as the number of solar farms on a system grows, pushing utility companies to seek novel solutions for power system protection. For example, it is known that fault currents coming from inverter-based solar photovoltaic plants are higher than fault currents from conventional generators. However, their time scales are much shorter, making traditional protection systems unsuitable for dealing with them.

6.5.1.4   Modelling of PV Plants for Grid Analysis

Utility companies tend to view distributed solar generation as net load, which implies that neither the utility company nor the system operator may be sufficiently aware of actual power generation or total local load at any point in time. This lack of information makes it very difficult to build an accurate model of the local load, which may serve for load forecasting, or the assessment of system reliability and security of the grid in the event of a failure of the solar plant.

6.5.1.5   Voltage Regulation

Photovoltaic solar farms directly influence voltage profiles along distribution feeders. This influence may potentially send voltages outside the allowed range, requiring voltage regulation mechanisms. In the absence of voltage regulation, this problem is exacerbated if the voltage profiles are not visible by system operators, and thus the deployment of appropriate monitoring systems becomes essential.

6.5.2   Criteria for Solar Power Integration

The connection to the power grid of variable renewable electricity generation, such as solar plants, requires the analysis of several factors which may impact the grid's operation. Note that the main criteria for integrating solar plants is dictated by the locally applicable standards and regulations, which vary from country to country. A major criterion for plant connection is the impact on the grid voltage during normal operations, such as slow voltage variations. The plant to be connected is required to keep the voltage increase in an acceptable range of typically 2–3%. The voltage increase depends on the network short-circuit power and the impedance angle at the connection point. It also relates to equipment characteristics, particularly transformers and distribution lines, and is affected by the phase angle of the plant, which can be adjusted by reactive power control.

Modern inverters can provide grid-supporting functions, such as frequency droop and reactive power support to help maintain local grid parameters within their required limits. Moreover, three important technical parameters affecting the quality of the power injected into the grid by inverters are harmonic distortion, flicker and DC injection. The applicable standards regulate these criteria, and thus inverter manufacturers need to ensure that their products comply with the appropriate limits in the markets where they are sold.

A further criterion for connecting a renewable electricity generation plant to the distribution grid is the thermal limit of the grid components (mainly electric power lines). These thermal limits are related to the short circuit levels, which are increased by the presence of generation in the vicinity of the components.

The scenario in which a grid-connected PV facility continues to energise the circuit even when utility grid power is unavailable is known as *islanding*. Utility personnel may not recognise that a circuit is still powered while working on repairs or maintenance, which might compromise their safety. As a result, most countries mandate anti-islanding capability, which requires the PV system to stop energising the grid when grid power is unavailable.

In conventional power grids, inverter-based distributed generation plants must typically be disconnected when the grid voltage or frequency exceeds the allowable operating range. This action is taken to avoid islanding, as discussed above. However, in power grids with a high share of distributed renewable generation, the simultaneous loss of many generation plants can threaten the grid's overall stability. The ability of generation plants to remain connected to the network during faults of short duration—also referred to as Fault Ride-Through (FRT) capability—is crucial for the integration of large-scale renewables into the power grid. As a result, modern inverters are provided with functionality that allows them to stay connected to the grid during short grid faults through the low/high voltage ride-through (LHVRT) and low/high-frequency ride-through (LHFRT) functions.

6.6 Approaches for Grid Integration

As the share of renewables in overall power generation is rapidly increasing in many countries, the structure and operation of the existing power grid need to be adapted to accommodate this increasing share of renewables. Utility companies and power system operators have to introduce new technologies and procedures in this adaptation process. This section discusses some of the options that utility companies have to facilitate the integration of renewables and the technologies employed for that purpose.

6.6.1 Options to Facilitate the Integration of Solar Power

Some options are available to make the integration of solar power and other variable renewables easier. These are discussed below.

- *Curtailing renewable generation.* As the installed capacity of renewable generation increases, the total generation level can exceed the demand. A relatively easy option to address this issue is to curtail excessive generation. For example, this could be done by shutting down or reducing the power generated by some solar farms. Curtailing variable renewables is also used to prevent the violation of thermal and voltage constraints in distribution networks. Frequent curtailment may be undesirable for the generator as it could represent a significant loss of revenue. However, some markets have contractual mechanisms that compensate for the generator's flexibility.

- *Using fast responding generators.* Natural gas plants can quickly change their generation level. Therefore, they can compensate for fluctuations in renewable power by adjusting their generation level as the renewable generation changes. The ability of these flexible generators to respond rapidly can be encouraged and compensated by financial means. However, gas power plants have a high carbon intensity compared with all renewable energy sources, and thus this is not the best solution for environmental reasons.

- *Using energy storage.* Energy storage, which is further discussed below in this chapter, is the capture of energy produced at one time for use later. Energy conversion is necessary for most energy storage devices, and efficiency levels vary between technologies. Some existing energy storage technologies used in power systems include batteries, flywheels, hydrogen storage, compressed air and pumped hydro.

- *Demand-side response.* Demand-side response is a scheme utility companies use to influence their customers' energy demand profile through automated or manual mechanisms. This approach helps shift flexible demand to times of high renewable generation (for example, heating water at times of high solar PV generation) and reducing power demand at peak hours.

- *Mixing different types of renewable sources.* This mixing results in the aggregation of variable power from different sources (e.g. wind + solar) with unrelated inherent causes for their variability, which can help reduce the intermittence of a single type of resource.

Most of the above approaches require the use of smart grid technologies for their effective implementation. For example, a comprehensive and modern communications and sensing infrastructure is required in most of them. Furthermore, demand-side response schemes rely on the smart metering infrastructure. Moreover, curtailing solar generation is often implemented in distribution networks via *active network management* (ANM), an approach that is discussed below.

6.6.2 Active Network Management

Active network management (ANM) is a control system that enables utility companies to manage distributed generation, storage and flexible demand in real-time to increase the utilisation of network assets without breaching operational limits, reducing the need for network reinforcement, facilitating connections, and reducing costs [7].

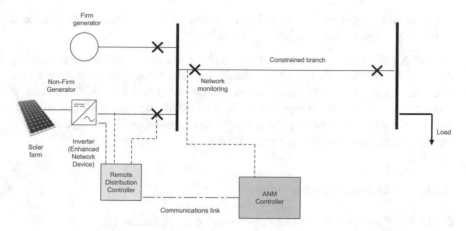

Fig. 6.12 Illustration of ANM system

There are three core aspects to ANM. Firstly, network monitoring data providing near real-time information on the state of the network. Secondly, a processing function to assess the network, given the monitoring data provided, and recommend adjustments to the network configuration to optimise performance and avoid voltage or thermal violations. The third aspect is connection to active network devices, which allows the system to issue setpoints or send commands to implement the network optimisation. See Fig. 6.12.

A key requirement of an ANM is for it to actively manage active network devices across a given area to control network constraints. The ANM system can then be parameterised for the priority it places on constraints and which devices it uses to avoid their violation.

The following factors act as drivers for the development and deployment of Active Network Management:

1. *Increasing distributed generation.* The increasing amount of small localised distributed generation sources on the network may result in complex bi-directional power flows in some branches.
2. *Low Carbon technologies.* Many governments have made commitments to minimise CO_2 emissions. This involves increasing the share of renewables in the generation mix and transitioning to electric heating and transportation, among other changes.
3. *More stochastic customer load profiles.* Passive distribution network management is appropriate when customer load profiles are predictable across periods ranging from days to weeks. However, this control philosophy breaks down when the profiles grow more stochastic, necessitating the use of more dynamic approaches.
4. *Preference by utility companies to avoid or defer network reinforcement.* As demand and renewable energy grow, distribution systems become increasingly

constrained, necessitating action by utility companies to extract headroom from existing infrastructure or reinforce networks.

The ANM technology allows loads and generators to connect to existing distribution networks that were previously thought to be full. It is hoped that grid connections should be faster and more cost-effective with ANM. The utility company can thus make better use of its assets and avoid or postpone network upgrades. ANM is well-suited to incremental development, offers network operating flexibility, and improves network performance.

6.6.2.1   Principles of Access

Principles of access through ANM can be one of the following [8]

- *Last-In-First-Out (LIFO)*, This approach priorities the oldest connections when deciding on curtailment based on current capacity. However, it is expandable such that new entrants will have access to the capacity as it becomes available. In cases where early projects require assurance and simplicity, this approach may be preferable.
- *Pro-rata*, where curtailment is shared across generators. This approach may be preferred where maximising connections is an objective and applications are reasonably concurrent.

6.6.2.2   Third-Party ANM

The third party ANM concept involves a system that is installed and managed by a customer behind the meter, as opposed to the utility company. The third-party ANM may provide local control for export limitation, it may manage the interconnection with the utility grid, or the integration point for distributed energy resources to a full ANM system. The system may also serve as the interface at the grid edge for scheduling resources, dispatch of generation, visibility (from the perspective of the utility company), and data collection. The third-party ANM may provide local control of constraints and objectives, or direct dispatch of generation. There are two types of third-party ANM connections, both of which are installed and managed by the customer.

- *Shared capacity*. This is the case when an existing generator may have a contracted capacity but only has a lower amount of connected generation. Therefore, another customer can approach this generator and make use of the spare capacity of the connection. The customers will install a system that will ensure that the combined export of both generators does not exceed the contracted capacity.
- *Demand management*. This is used when an existing generator with a limited export contract increases its installed demand to the point that, when the

generator exceeds its contracted limit, additional demand is activated to avoid the breach.

The utility company will typically install a backup system so that when the customer's system fails, the generators are disconnected.

6.6.2.3 Communications Networks for ANM

Full ANM installations typically use satellite links for the required data connections between the different elements of the ANM system, which can be distant. This type of connection can be costly, but it would not be necessary for a behind-the-meter third-party ANM deployment, as relative physical proximity of the different elements of the ANM system is most likely in this case. It may be the case that the utility company will need to approve the performance, reliability and security of the communications network in a third-party ANM system. However, it is quite possible that fibre-optic connections or some form of secure wireless network will be both acceptable for the utility and cost-effective for the customer.

Wireless mesh networks are an example of potentially suitable wireless communications technology that could be used in the case of a third-party ANM deployment. These networks are composed of cooperating radio nodes organised in a mesh topology. Wireless mesh networks offer ease of operation and redundancy thanks to their mesh properties, self-configuration and self-healing properties [1].

6.6.3 Other Flexible Options for Connection

In addition to ANM, potential customers wishing to establish new generator connections to the distribution networks may typically be offered by the utility company some of the following flexible connection options [9]:

- *Single Generator Active Network Management (SGANM).* This connection option is identical to a full ANM scheme, however it only manages one generator and up to two constraints instead of multiple constraints and multiple generators.
- *Timed connections.* The timed export connection gives customers the option of connecting to the network and exporting at specific times. It is possible to exploit generation or demand diversity in specific sections of the network.
- *Export limiting devices.* An export limiting device measures the apparent power at the grid connection point and utilises that information to limit power generation or balance customer demand to avoid exceeding the contractual export capacity.
- *Intertrip.* The utility company must normally ensure that if there are two circuits in parallel and one of the circuits fails, then the other circuit must be able to absorb the load. A generator that applies for a connection may incur reinforcing charges if the remaining circuit does not have sufficient excess capacity. An

intertrip connection may be offered by the utility company in this case, which allows the customer to connect to the network such that the generator will be disconnected if one of the circuits fails.

6.7 Energy Storage and PV Integration

Energy storage can assist PV integration by increasing power system flexibility. Many energy storage technologies are available, and most of the existing utility-scale storage capacity involves pumped hydro storage facilities [2]. However, the cost of battery-based energy storage has gone down significantly in recent years, which has stimulated interest in deploying battery technology in power systems [10]. Moreover, pumped storage does not have the agility to compensate for the fast patterns of variability of intermittent renewables. Using suitable energy storage systems, the profile of PV production can be adjusted to better match load and peak demand in the power grid. This possibility can help make PV generation more flexible and facilitate higher PV penetration levels in the power grid.

Battery storage benefits PV installations of different scales, from small residential PV installations to large solar farms. Figure 6.13 shows a grid-scale 2 MWh battery storage container, which is co-located with a 7.4 MWp photovoltaic array at Westhampnett solar farm in West Sussex, England.

Battery storage systems can be physically sited at various locations within a power grid. Storage does not need to be co-located with PV or other variable renewables to produce benefits. There are different categories of association between energy storage and photovoltaic plants. These are discussed below [11].

- *Independent*—PV and battery are not co-located and do not have a point of common coupling (PCC) with the grid; the energy stored in the battery could come from either PV or the grid.
- *AC-coupled*—PV and battery are co-located and have a common point of common coupling with the grid, as illustrated in Fig. 6.14. The energy stored in the battery could come from either the grid or PV. In an AC-coupled system, the PV energy is converted to AC, and converted back to DC to store it in the battery. When the battery discharges, the energy is converted back to AC. Separate inverters are required in this case for the solar array and battery storage. With every conversion, some energy is lost. However, with the efficiency of modern inverters, this loss is likely to be relatively small.
- *DC-coupled*—the battery is connected to DC side of PV inverters, as illustrated in Fig. 6.15. In this case, the energy stored in the battery could come from either the main grid or the PV array. In DC-coupled PV systems, the PV energy stored in the battery is not converted to AC and back to DC. Therefore, only one inverter is required in this case, facilitating installation, reducing hardware costs, and making the whole PV system more economical and efficient, compared with the AC-coupled case.

Fig. 6.13 Grid-scale 2 MWh Lithium-Ion battery storage container at Westhampnett solar farm in West Sussex, England. The total battery storage capacity in this solar farm is 4 MWh, which is co-located with a 7.4 MWp photovoltaic array

- *Tightly DC-coupled*—the battery is connected to the DC side of PV inverters; charging is controlled in such a way that energy stored in battery can come only from PV.

AC and DC-coupled PV and storage are often used to maximise self-consumption of solar energy at sites where there is also a local electricity demand behind the point of common coupling. This application is attractive when exporting electricity to the power grid brings little benefit due to low export prices. Another benefit of co-located systems is that the storage can often provide a backup service in case of power outages affecting the grid.

6.8 Microgrids and Solar Energy

A microgrid is a local energy system consisting of a grouping of small scale generating units, loads and possibly energy storage, all within a bounded and controlled network, and which may or may not be connected to the grid [12].

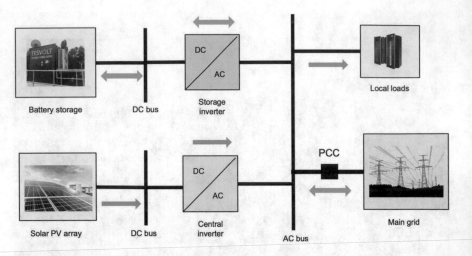

Fig. 6.14 AC-coupled PV and storage. The green arrows indicate the possible directions of power flows. PCC is the point of common coupling with the grid

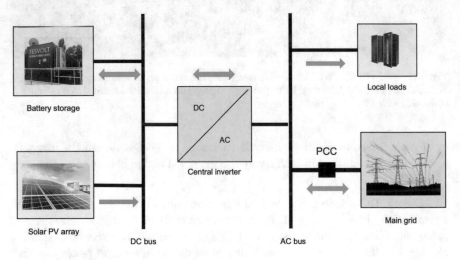

Fig. 6.15 DC-coupled PV and storage. The green arrows indicate the possible directions of power flows. PCC is the point of common coupling with the grid

The microgrid model offers numerous advantages to those who adopt it. First, microgrids enable grid modernisation by facilitating the integration of multiple smart grid technologies. Second, microgrids are known to enhance the integration of distributed and renewable energy sources, including solar energy, into the main power grid. Third, they promote energy efficiency and reduce losses by locating generation near the location of the demand. Fourth, they help improve reliability and power quality.

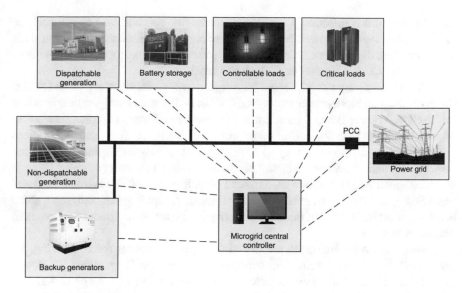

Fig. 6.16 Microgrid components

Microgrids have the potential to reduce significant capital investments by meeting increased consumption with locally generated power. They meet end-user needs, for example, by securing energy supply to critical loads. Microgrids also promote participation in new markets, including demand-side management, load levelling and ancillary services (see Sect. 6.9). They promote community energy independence. Microgrids also support the primary grid and enhance the integration of intermittent renewables. Finally, they have the potential to reduce the carbon footprint by maximising clean local generation.

Microgrids can be classified as remote and off-grid microgrids, commercial and industrial microgrids, community microgrids, mission critical microgrids (such as data centres), and institutional microgrids (such as hospitals and university campuses).

The components of microgrids usually include distributed energy resources, control and management subsystems, secure network and communications infrastructure, and a point of common coupling (PCC), as illustrated in Fig. 6.16. When renewable energy resources are included, they usually are of the form of relatively small wind or solar generation plants, waste-to-energy plants, and combined heat and power plants (CHP).

6.8.1 The Point of Common Coupling

The point of common coupling (PCC) is where a microgrid connects to the main grid. Circuit breakers, power electronic interfaces, and static switches can all be used to make this connection. Circuit breakers are low-cost and easy-to-use devices. They are, however, slow, needing 3–6 cycles to complete a disconnect. On both sides of the breaker, the network type must be the same. Direct control of the power flow through the PCC is not possible with circuit breakers.

Static switches are usually based on semiconductor controlled rectifiers (SCRs) or insulated gate bipolar transistors (IGBTs), in antiparallel configuration to allow bidirectional power flow. They are more expensive and more complex than using circuit breakers. Usually, conventional circuit breakers are still used in series to static switches to provide a way to achieve complete galvanic isolation. They act much faster than conventional circuit breakers, usually in the order of half a cycle to a cycle. Sometimes IGBTs are used instead of SCR because IGBTs tend to be faster than SCRs, and their current is inherently limited. The power flow through a static switch cannot be controlled directly. Moreover, there are some conduction losses in these devices.

Power electronic interfaces are the most expensive option for connecting the microgrid to the main grid, but they are also flexible. They allow for power distribution architecture characteristics on both sides of the PCC to be completely different (for instance, AC and DC). Both real and reactive power flow can be controlled. Reaction times to connection or disconnection commands are similar to those provided by static switches. Still, in many cases, a circuit breaker is required at the grid-side terminal of the power electronic interface to provide a way to disconnect the microgrid from the grid physically. The presence of a power electronic circuit will lead to some power losses.

6.8.2 Microgrid Control, Monitoring and Optimisation

The microgrid control system monitors, automates and optimises the use of distributed energy resources within the microgrid [13]. The monitoring function frequently checks the health of the microgrid's operation and its different components. It also detects faults and handles alarms. The microgrid control involves frequency and voltage control, reserve management, grid restoration, black start sequencing, load and generation control, load shedding, and other functions. Optimising a microgrid involves coordinating the timing, selection, and generation levels of dispatchable distributed energy resources with non-dispatchable ones to minimise energy costs.

The microsource controller (MC) monitors and controls distributed energy resources, such as distributed generation, storage, and loads. The microgrid central controller (MGCC), on the other hand, provides the main interface between the microgrid and other actors (such as the utility company) and can assume different roles, including the optimisation of the microgrid operation and the coordination of local microsource controllers.

Microgrids can be operated in two different modes interconnected and islanded. The connection with the main grid is active in interconnected mode, and in this case the microgrid's generators supply, at least partially, the loads, imports power from the grid, or exports power to the grid. The main grid gives the frequency in this mode, and it has a significant impact on voltage levels across the microgrid. The microgrid central controller communicates with the several microsource controllers as well as the distribution management system of the utility company. The output control of dispatchable generators is on dispatch power mode when the microgrid is in interconnected mode, therefore the output of dispatchable generators is determined by economic dispatch calculations. On the other hand, the microgrid is disconnected from the main grid when it operates in islanded mode. In this case, the microgrid central controller switches the output control of diesel AC generators from dispatch to frequency mode in this scenario, such that the islanded microgrid's frequency can then be precisely controlled.

6.8.3 Microgrid Communications Infrastructure

Some key microgrid components, including distributed energy resources and loads, need to be equipped with communication capabilities that enable them to exchange data over communication links. Intelligent electronic devices (IEDs), which enable this functionality, can communicate between them, and also with the microgrid central controller.

The communication infrastructure has an important role in the control and operation of the microgrid. Although many microgrids still use older communication technologies, new standards and protocols are being developed to address the particular challenges and requirements of microgrids.

The communication network can be organised into different area networks based on the operation levels including home area network (HAN), building area network (BAN), industrial area network (IAN), field/neighbourhood area network (FAN/NAN), and wide area network (WAN).

A communication link can employ wired or wireless communication media. Wired technologies provide high reliability and security but are more expensive to install and lack flexibility to network changes, compared to wireless networks. Wireless technologies may suffer from environmental interference and transmission attenuation. On the other hand, they are relatively cheap to install, are easy to scale up, and are much more flexible than wired networks.

For communicating measurements and control signals at the device level, wired technologies such as cables and optical fibres are often employed. Power line communications (PLC) can be used to establish links between the secondary and tertiary controllers. HAN usually employs wireless technologies such as Bluetooth, ZigBee, and Wi-Fi. For FAN/NAN technologies such as PLC, DSL, cellular communication and fibre-optic cables are often employed. The WAN network may use fiber-optic cables, cellular communications, or satellite links.

6.9 Selling Solar Electricity and Market Opportunities

The conjunction of smart grid technologies and modern deregulated electricity markets open a range of opportunities for owners of solar generation facilities of different sizes to participate in, and profit from, various electricity markets. Some of these opportunities are discussed below.

6.9.1 Power Purchase Agreements

Power purchase agreements (PPAs) are long-term renewable energy contracts. In PPAs both parties agree on various aspects, such as the price of energy, amount of energy delivery, and duration of the contract.

On-site PPAs involve renewable energy installations located on a client's premises, for instance on rooftops, or adjacent land. In this case, the energy provider funds the design, installation and operation of the equipment. The client consumes the energy that is produced, and pays an agreed price for it.

Off-site PPAs, on the other hand, involve renewable energy installations that are not located on a client's premises. Here, the client agrees to buy a quantity of energy produced by a renewable energy facility owned by an energy provider. The connection between the client and the renewable energy facility can be direct and physical by means of a private wire. The connection can also be indirect via the power distribution network, through an arrangement known as *sleeving*. For example, the energy seller could be a solar farm owner, and the energy buyer could be a utility company, a private company, or a local authority.

6.9.2 Export Tariffs and Net Metering

A simple way for owners of photovoltaic assets to participate in the market may be to sell excess electricity back to the grid. The possibility of doing so depends on the local regulations, but where this is possible an export tariff per kWh of electricity is typically paid to the customer. Smart meters usually support the metering of exported electricity. This can be exploited at different scales, from households to solar farms. Unfortunately, governments are prone to changing the renewable energy policies and these tariffs tend to change or even disappear. In the UK, for example, *feed-in tariffs* were available between 2010 and 2019 to encourage, via tax-free payments under a contract with a duration of up to 25 years, the installation of renewable energy generation with capacities of up to 5 MW. At the beginning of the scheme, the tariffs were very attractive and resulted in a measurable increase in photovoltaic and other renewable energy installations. However, the progressive reduction in the level of the feed-in tariffs and other changes meant that, over time, the scheme became less and less attractive until it was closed to new applicants in 2019. From 2020, the UK government established the Smart Export Guarantee scheme, which requires electricity suppliers to pay small scale generators (up to 5 MW) for low-carbon electricity. The tariffs that are paid under this new scheme, however, tend to be only a small fraction (as low as 10%) of the price electricity is sold to customers.

In some regions, solar photovoltaic system owners can be credited for the electricity they add to the grid through a scheme called *net metering*. For example, consider a PV system on a residential customer's roof which produces excess electricity during the daytime. If the home is net-metered, the electricity meter will run backwards to provide a credit against the amount of electricity used at night or during other times when the home's electricity consumption exceeds the system's output. Customers are only charged for the net amount of energy they use. Differences between regional legislation, regulatory decisions and implementation policies cause the mechanism for compensating customers who are solar producers to vary widely.

6.9.3 Virtual Power Plants and Electricity Markets

The term virtual power plant (VPP) is used to describe a collection of power generation sources, energy storage devices and demand response participants which are spread within a distribution grid [14]. Distributed energy resources can be attractive for investors. Nonetheless, their intermittency makes it difficult to dispatch these resources. Combining them into a larger VPP that can be dispatched and controlled efficiently addresses this issue. Virtual power plants are acknowledged as a viable option for aggregating and operating distributed energy resources in order to participate in wholesale energy markets while also providing the flexibility and grid services required in a renewable-rich energy system.

For example, the Tesla South Australian VPP project is connecting thousands of homes equipped with PV arrays and smart batteries into a virtual power plant that, as of March 2021, had a capacity of 13 MW [15], and its usage includes wholesale market arbitrage, as well as participation in the frequency and ancillary services market.

6.9.3.1 Electricity Markets

In many countries or regions, electrical energy is generally traded in large quantities in an electricity market [16]. Electricity markets usually involve different trading types, including the day-ahead, the real-time, and the futures markets [17]. The day-ahead market allows trading electricity commodities once a day, one day in advance, and on an hourly basis. On the other hand, the real-time market, also known as the balancing market, closes some minutes before the actual power delivery. In addition, the futures market allows transactions from 1 week to several years in advance [18].

6.9.3.2 Futures Market

Energy transactions via futures markets are possible with VPPs. The risks associated with the significant price unpredictability in the day-ahead electricity market by participating in this market. To boost their operating profit, VPPs take advantage of arbitrage opportunities between the day-ahead electricity market and the futures market.

6.9.3.3 Day-Ahead Market

The day-ahead market uses bids from market agents to perform electricity transactions for each hour of the following day. The VPPs give generators direct access to electricity markets for selling of their generated energy, as well as allowing customers who also generate energy to sell excess energy.

6.9.3.4 Ancillary Services Market

The primary goal of the ancillary services market is to ensure the security and reliability of the electricity grid by providing the capacity for the system to maintain a good balance between generation and demand. VPPs can participate in ancillary services markets by providing frequency response and backup power services, supporting the quality and security of the electricity supply.

6.9.3.5 Intraday Market

As additional information is available during the day, intraday markets are designed to make adjustments to the energy exchanged in the day-ahead market. The volume of energy exchanged in intraday markets is lower than in the day-ahead power market. Because of the rise in renewable energy and its unpredictable character, intraday markets are becoming more important, making it necessary to change offers and fix the imbalances in predicted generation availability. VPPs can invest in intraday markets to boost their returns.

6.9.3.6 Real-Time Balancing Market

The last opportunity for balancing generation and demand is known as the real-time balancing market. Here, transactions close somewhere between 5 and 30 min before real energy delivery. Despite the fact that intraday markets allow VPPs to adjust scheduled energy after the day-ahead market, exchange power imbalances may still emerge when dispatch time approaches. VPPs can thus participate in real-time balancing markets to avoid penalties.

6.9.4   Microgrids and Energy Markets

Local energy markets based on microgrids have been proposed [19]. Consumers and prosumers (those that both generate and consume energy) can trade energy with one another in this peer-to-peer marketplace. As a result, consumers and prosumers can keep the income from energy trading in their communities. This encourages investments in renewable energy plants and helps to balance supply and demand locally. As a result, both financial and socioeconomic incentives are offered for the integration and spread of locally produced renewable energy. Innovative information solutions for integrating market players in a user-friendly and comprehensive manner are required for the smooth running of these microgrid-based energy markets. Blockchain-based technology has been proposed to build a local energy market without the involvement of central intermediaries [20]. A blockchain is a decentralised database that is shared among computer network nodes [21]. Blockchains are well known for their critical role in keeping a secure and decentralised record of transactions in cryptocurrency systems like Bitcoin. The blockchain's novelty is that it ensures the fidelity and security of a data record while also generating trust without the requirement for a trusted third party.

Microgrids have also been proposed to provide ancillary services [22]. Microgrids can supply some of these services and as a result earn revenue from the utility company. Consequently, a grid-connected microgrid can help strengthen the utility grid.

References

1. Stephen F. Bush. *Smart Grid: Communication-Enabled Intelligence for the Electric Power Grid*. Wiley, 2014.
2. Shady S. Refaat, Omar Ellabban, Sertac Bayhan, Haitham Abu-Rub, Frede Blaabjerg, and Miroslav M. Begovic. *Smart Grid and Enabling Technologies*. IEEE Press and Wiley, 2021.
3. Bernd M. Buchholz and Zbigniew Styczynski. *Smart Grids – Fundamentals and Technologies in Electricity Networks*. Springer, 2014.
4. Geoff Stapleton and Susan Neill. *Grid-connected Solar Electric Systems*. Earthscan, 2012.
5. ABB Inc. Integrating small solar farms to the grid: a 'smart' guide.
6. University of Sheffield. PV Live website.
7. Quentin Gemine, Damien Ernst, and Bertrand Cornélusse. Active network management for electrical distribution systems: problem formulation, benchmark, and approximate solution. *Optimization and Engineering*, 18(3):587–629, 2017.

8. Laura Kane and Graham Ault. A review and analysis of renewable energy curtailment schemes and principles of access: Transitioning towards business as usual. *Energy Policy*, 72:67–77, 2014.
9. Scottish and Southern Electricity Networks. Flexible Connections.
10. Will Gorman, Andrew Mills, Mark Bolinger, Ryan Wiser, Nikita G. Singhal, Erik Ela, and Eric O'Shaughnessy. Motivations and options for deploying hybrid generator-plus-battery projects within the bulk power system. *The Electricity Journal*, 33(5):106739, 2020.
11. Gevorgian Vahan, Robb Wallen, Przemyslaw Koralewicz, Emanuel Mendiola, Shahil Shah, and Mahesh Morjaria. Provision of grid services by PV plants with integrated battery energy storage system, 2020.
12. S.P. Chowdhury S. Chowdhury and P. Crossley. *Microgrids and Active Distribution Networks*. The Institution of Engineering and Technology, 2009.
13. Hassan Bevrani, Bruno Francois, and Toshifumi Ise. *Microgrids Dynamics Control*. Wiley, 2017.
14. Rakshith Subramanya, Matti Yli-Ojanperä, Seppo Sierla, Taneli Hölttä, Jori Valtakari, and Valeriy Vyatkin. A virtual power plant solution for aggregating photovoltaic systems and other distributed energy resources for northern european primary frequency reserves. *Energies*, 14(5):1242, Feb 2021.
15. SA Power Networks. Advanced VPP Grid Integration, 2021.
16. Deqiang Gan, Donghan Feng, and Jun Xie. *Electricity Markets and Power System Economics*. CRC Press, 2014.
17. L. Baringo and M. Rahimiyan. *Virtual Power Plants and Electricity Markets: Decision Making Under Uncertainty*. Springer, 2020.
18. Natalia Naval and Jose M. Yusta. Virtual power plant models and electricity markets - a review. *Renewable and Sustainable Energy Reviews*, 149:111393, 2021.
19. Esther Mengelkamp, Johannes Gärttner, Kerstin Rock, Scott Kessler, Lawrence Orsini, and Christof Weinhardt. Designing microgrid energy markets: A case study: The Brooklyn microgrid. *Applied Energy*, 210:870–880, 2018.
20. D. Strepparava, L. Nespoli, E. Kapassa, M. Touloupou, L. Katelaris, and V. Medici. Deployment and analysis of a blockchain-based local energy market. *Energy Reports*, 8:99–113, 2022.
21. Andreas Bolfing. *Introduction to Blockchain Technology*, chapter 6. Oxford University Press, 2020.
22. Alireza Majzoobi and Amin Khodaei. Application of microgrids in providing ancillary services to the utility grid. *Energy*, 123:555–563, 2017.

Chapter 7
Applications of Solar Energy

7.1 Rooftop Solar Installations

Rooftop solar installations are being increasingly deployed in many countries. Solar panels are typically retrofitted on existing roofs, but many new buildings are being constructed with solar panels already installed.

Rooftop solar does not require extra space for installations, as existing rooftop space is used. Apart from reducing their owner's carbon footprint, collectively, rooftop installations reduce a country's dependency on fossil fuels and their harmful emissions.

The significant potential of rooftop solar energy is illustrated in a study by the National Renewable Energy Laboratory (NREL) in 2016 [1], which concluded that in the US there are more than 8 billion square meters of rooftops on which solar panels could be installed, which represent over 1000 GW of potential solar capacity. For example, this potential solar capacity is about 5 times higher than the peak power demand in the US recorded on 1st April, 2022, which was 198.47 GW.

Rooftop solar installations help reduce energy bills. For example, rooftop photovoltaic panels supply electricity to buildings, so they need to import less electricity from the grid, resulting in savings in the electricity bill. Moreover, thermal solar rooftop installations help reduce fuel bills associated with hot water, which is often produced by boilers that burn natural gas.

Rooftop photovoltaic installations usually represent a secure investment, with payback periods typically in the range of 6–8 years. Given that solar panels will last for 25 years or more, they produce positive financial returns over time. In developing countries with limited electrification, rooftop photovoltaic installations help expand access to electricity to households that are not served by the electricity grid.

© The Author(s), under exclusive license to Springer Nature Switzerland AG 2023
A. Rachid et al., *Solar Energy Engineering and Applications*, Power Systems,
https://doi.org/10.1007/978-3-031-20830-0_7

7.1.1   Factors that Influence Rooftop Installations

A number of key factors that influence rooftop solar installations are discussed below.

- **Type of roof** Roofs can be broadly classified according to their slope. Roofs are considered to be flat if they have a slope that is lower than 5° with respect to the horizontal, slightly sloped if the slope is between 5° and 22°, normally sloped if the slope is between 22° and 45°, and steep if the slope is higher than 45°. However, there are many other types of roof shapes, some of which have sections with different slopes, and roofs that are curved. The module mounting system that is selected for rooftop solar energy installations depends on the roof type and the structural properties of the building.
- **Roof orientation** In the case of sloping roofs, roof orientation influences the location of the installation on the roof, with south-oriented roofs being the best orientation in the northern hemisphere. Flat roofs allow the possibility of adjusting the orientation of the solar panels to maximise the capture of solar energy.
- **Roof covering** These are individual components such as clay tiles, stone slabs, slates, shingles, and roofing sheets that are utilised for drainage purposes in sloping roofs.
- **Roof sealing** This is an essential component of flat roofs that consists of a fully waterproof layer that covers the entire surface of the roof and is constructed of materials such as bitumen roofing felt, plastic roof sheeting, or plastics applied as a liquid and later hardening.
- **Superstructures** Roofs may also have superstructures such as chimneys, skylights, and dormer windows. However, these fixtures and constructions can diminish the available roof space for installing solar arrays and project undesirable shadows on solar panels.
- **Structural integrity** A qualified professional should evaluate the structural integrity and condition of the roof and building to ensure that the additional load caused by the solar array and the mounting structure does not exceed the building's authorised loading limitations.
- **Wind and snow loading** The mounting system should withstand the locally expected wind and, if appropriate, snow loading. Therefore, the choice of the mounting system and the calculation of design parameters (such as the weight of ballasts), should be made considering such loadings.
- **Shading from nearby structures** Shading from other buildings or objects, such as trees, affect the capture of solar energy, and influences the location of the solar installation on the roof.
- **Health and safety** The system's physical layout should consider local health and safety standards, such as whether emergency services need access in the event of a fire. These considerations also impact system maintenance tasks.

7.1.2 Mounting Systems for Sloping Roofs

Mounting systems for sloped roofs can be classified as on-roof and in-roof [2]. These systems are discussed below.

7.1.2.1 On-Roof Mounting

In the case of on-roof mounting, the modules are fitted above the existing roof covering using a metal structure, as illustrated in Fig. 7.1. The roof covering is kept in place and continues to provide water resistance. On-roof mounting is usually the most cost-effective solution for installing new arrays to existing roofs.

The modules should be positioned so that a self-contained, rather than disjointed, PV array surface is created, allowing the system to blend in as seamlessly as possible with the existing roof surface.

Roof mounts, mounting rails, and module fixings are the three primary components of the metal structure that supports the modules. Firstly, a rail system attaches to the roof structure beneath the roof covering or secures directly to the covering using roof mounts. Secondly, module fixings anchor the modules to the rails.

Fig. 7.1 Example of an on-roof photovoltaic installation on a sloping roof

Fig. 7.2 Illustration of an in-roof solar system at a car park at the Autonomous University of Madrid (UAM), Spain. Source: image by Hanjin is licensed under CC BY-SA 3.0

7.1.2.2   In-Roof Systems

The modules for in-roof mounting lie in the plane of the regular roof covering. In-roof modules can cover the entire roof surface or just a portion. Here, the photovoltaic array serves the dual purpose of power generation and weatherproofing. As a result, the mounting system between the modules and at the array's borders must be fully weatherproof. In addition, sufficient ventilation behind the modules is needed to prevent moisture damage to the roof caused by condensation. Figure 7.2 illustrates a kind of in-roof solar system used in car park canopies.

7.1.2.3   Solar Tiles

There are two types of roof tile. One type of roof tile comprises solar modules that have been adapted to be comparable to traditional roof coverings; they are large enough to replace multiple roof tiles while reducing the amount of electrical wiring. Usually, these roof tiles can be attached directly to the existing roof battens.

The other type of roof tile is made up of pieces of roofing material that have a solar module attached to them at the factory. The roofing material provides mechanical support and weatherproofing for the embedded PV module. These roof tiles are often relatively small, which has the advantage of allowing them to be installed on complex roof surfaces, as well as on roofs of historic structures. A group of houses equipped with solar tiles on their roofs is illustrated in Fig. 7.3.

Fig. 7.3 Illustration of a group of houses with solar tiles on their roofs. Source: 'Solar tiles on house roofs' by Stephen Craven is licensed under CC BY-SA 2.0

7.1.3 Mounting Systems for Flat Roofs

Flat roof modules are mounted on a metal structure atop the existing roof surface, similar to roof-mounted systems on sloping roofs [2]. The modules are typically tilted at a suitable angle by the support structure.

The method of securing mounts to flat roofs needs to be chosen carefully. Expected wind forces must be taken into account when securing arrays. The choice of fixing depends upon the structure of the roof, while the capacity of the roof to accept significant additional loads determines whether the system can be free-standing or must be anchored to the roof. In case of profiled sheet roofs, the mounting frame is fixed directly to the roof covering.

With ballast-mounted systems, the flat roof mounts are anchored without penetrating the roof. For example, purpose made plastic mounts are available that can be filled with ballast and used to install solar panels with a tilt angle, as illustrated in Fig. 7.4. Alternatively, concrete slabs are often placed on the flat roof without any further fixing and the support frames are secured to these. Figure 7.5 illustrates the use of concrete slabs together with a support frame for the installation of solar panels on a flat roof.

Fig. 7.4 Illustration of the use of plastic mounts filled with ballast used on the flat roof of a Portsmouth City Council building, UK. Solar PV panels are securely anchored to the plastic structures that provide them with an adequate tilt

The main benefit of ballast-mounted systems is that they do not require roof penetration. The PV array and ballast, on the other hand, must be sufficiently heavy to keep the system firmly in place even when subjected to the maximum projected wind load. It is worth indicating that, as with other rooftop solar energy systems, it is critical to adhere to all applicable building laws and regulations in this case.

7.2 Building Integrated Solar PV Systems

Building Integrated Photovoltaics (BIPV) are systems where PVs are utilised as a part of the building architecture. BIPVs can be implemented in domestic and industrial constructions whereby conventional elements of the building envelope are replaced by PV modules. In addition to power generation, application of BIPV could lead to reduced cost of building materials. Further advantages include flexibility of BIPV as stand-alone or grid-connected systems. Consumption of generated electricity on site typically coincides with the peak demand and in this case transmission and distribution losses can be avoided. Alternatively, especially if

Fig. 7.5 Illustration of the use of concrete slabs together with a support frame for the installation of solar panels on a flat roof at New Carrollton Federal Office Building, New Carrollton, Maryland

storage is employed, electricity exported to the grid can be effectively utilised for peak shaving (Fig. 7.6).

BIPVs have steadily been gaining popularity as they are largely seen as a promising technology for achieving Net-Zero goals and especially important for Net-Zero buildings. Cost, reliability, efficiency of BIPVs have been extensively studied and play important roles when selecting these technologies. However, socio-cultural factors and building aesthetics are equally important [3].

Unlike Building Applied Photovoltaics (BAPV), where PV panels are installed on an existing structure, BIPVs effectively replace conventional building materials, for example: windows, parts of or the whole façade, tiles and roofs. Where BAPV can be added to various types of dwellings, including retrofitting, BIPVs are typically considered for new builds. Use of BIPVs is highly standardised with number of new and regularly updated ISO and IEC regulations which must be adhered to.

BIPVs can be categorised in several ways. Current market products are predominantly either thick silicon crystal-based products and thin film technologies, often on a glass substrate. This film technologies generally yield lower efficiencies, they are less adversely affected by high temperatures and are often used on vertical and curved surfaces. Various BIPV modules, tiles, foils and cell glazings are known by their market names. In terms of application, BIPV can feature on the building roof

or building façade, including walls, windows and shading devices. In 2021, 80% of the BIPVs were rooftop installation, with remaining 20% façade mounted [4].

Façade mounted BIPV may take form of PV panels being used for shading, as standalone systems positioned on the surface to create a shade or integrated into the building. Façade based BIPVs can be used as part of the building exterior wall, or integrated into windows as curtains. Conventional PV systems are blue or black while thin films are black or brown. To change the colour of the façade, lamination can be used to achieve different colours. However, lamination leads to reduced BIPV performance.

In BAPV, panels are mounted on the roof. Unsurprisingly, the most common BIPVs application is roof top mounted where solar panels are replacing a section of roof tiles. Due to seamless fitting of panels into the roof, this design is sometimes preferred for aesthetic reasons, even though due to the lack of air-cooling panel output decreases. An alternative is to cover the whole roof area with panels, so called complete solar roof. More recently solar tiles, imitating the shapes and colours of conventional roof tiles, are also available. In BIPV context, conventional glazing is replaced with solar glass. Transparency can vary. Use of solar glass can reduce the glare and act as thermal insulation. BIPV materials can act as noise insulators as well.

There is a number of commercially software tools available for modelling of electrical, thermal and solar, and optical performance assessment of BIPV systems [5]; Homer, TRNSYS, PVsyst, EnergyPlus to name a few. In addition to technical operation, software tools consider the location of the building as well as the climate, usage, occupant comfort and other factors. Advanced analysis is required before BIPVs are employed. Despite numerous benefits, there could be some adverse impact on the building (for example overheating) if building design and performance considerations have been overlooked. Aesthetics of BIPV has been much debated over the years and social acceptance has now been reached in many parts of the world.

7.3   Solar Vehicles (Cars, Boats and Planes)

The first solar car was invented in 1955. Its top was covered with 12 Selenium PV cells and it had a small electric motor rotating the rear wheel shaft (Sunmobile, W.G. Cobb, 31.08.1955). Although the first competition on the solar racing car was in 1985 (Tour Del Sol), solar car challenges became popular with the Australian World Solar Challenge (WSC) that has been organized first in 1987. The first WSC was organized by Hans Tholstrup, who is a Danish-born adventurer and traveled from Perth to Sydney (4130 km) in 20 days with a solar car called Quiet Achiever [6]. Nowadays, different concepts of solar car races are being organized in different continents. These challenges aims to push the boundaries of electric vehicles that generating its energy need from the sun while enabling the technology meet with the public. International Solar Car Federation regulated challenges in three classes: Challenger, cruiser and adventure. Challenger is the class that most technological one seater vehicles participate. Before 2019, these vehicles can also use GaAs PVs. After Bridgestone World Solar Challenge 2019, it is limited to use only silicon types of PV cells. The challenger class vehicles also have a battery pack with a capacity of 5 kWh mostly of LiIon type due to its energy density. Efficiency of each unit on challenger class solar car is crucial. Thus, each unit including, motor, motor driver, MPPTs, energy management system, telemetry, lights and drivers on these vehicles are main focuses of researchers. In Fig. 7.7, the temperature distribution of a challenger class solar car during a speed of 30 km/h, $G = 700 \, W/m^2$ and ambiance temperature is 25 °C where colours represent a scale from 27 °C to 65 °C [7]. This analysis is very important to calculate the energy which can be generated from solar array during travel of the vehicle because PV cells efficiencies are decreasing with increasing temperature. Since PV cells are connected in two or in three regions, efficiency change in any cell connected in series reduces the related region efficiency. In high speeds, this is less important since the air cools the surface homogeneously.

Solar cars can be near future solutions if the costs of technological parts like efficient solar panels, hub motors, motor drivers and LiIon batteries decrease. However, even with these disadvantages, in 2019 June, a company from Netherland,

Fig. 7.7 (**a**) Solaris 10 during Bridgestone World Solar Challenge 2019 and (**b**) Temperature distribution of the Solaris 8 Solar Car (V = 30 km/h, T_a = 25°C, G = 700 W/m2). Colors represent a scale from 27°C to 65°C

Fig. 7.8 Three of the solar cars in market in year 2021. (**a**) Aptera 2 Series, US. (**b**) Lightyear One, EU. (**c**) Sion–Sono Motors, EU

revealed a solar car as a product to market with a price of approximately 150 k€. The car is called as light year one. The car has 5 m² photovoltaics, 725 km of range and 83 kW/km of energy consumption with its efficient aerodynamics design.

In the USA market, Aptera 2 series was one of the solar cars in 2021. It is a two-seater solar EV. It has 700 Wp photovoltaics on top that provide approximately 75 km per day with the car's unusual aerodynamic design. It's a three wheeler and the shape is more like a small light airplane without wings as can be seen in Fig. 7.8. The drag coefficient (Cd) of Aptera is 0.13 whereas Tesla Model 3 has a drag coefficient of 0.23 and a new generation Toyota Corolla has 0.29. It can be charged also from the grid but efficient vehicle dynamics and self cooling feature of PVs make Aptera can travel a range of 1640 km with a single charge.

The Sono Motors Company's solar car Sion is a low budget four seater. It has 248 monocrystalline PV cells on the body of the car that provide a range of 112–245 km. It can be charged also from the grid. Without solar energy range extension, it has a range of up to 305 km with a single charge. The battery capacity of Sion is 54 kWh, the power of its motor is 120 kW and the total weight of the vehicle is around 1400 kg.

Solar energy can be implemented on different range of boats and using solar energy on marine vessels. In the last 25 years, it became popular to add PVs in yachts, pleasure boats and even in commercial ships. The high percentage of internal

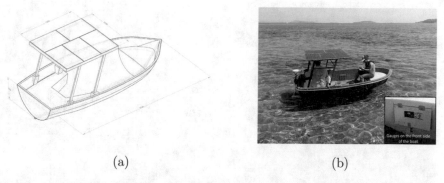

(a)   (b)

Fig. 7.9 A simple solar boat (drawing and photo by A.Gören). (**a**) Dimensions of a small fishing boat example. (**b**) A solar powered small fishing boat

combustion engine use on boats causes environmental pollution. Besides, the use of fossil fuels increases economic concerns of fishermen. Since energy consumption optimizations can be based on small or modular solutions more easily, small fishing or touristic boats are good choices to implement solar energy. As an example, Fig. 7.9 shows an experimental small solar boat which has been constructed in Dokuz Eylul University for researches on fisherman boats that are very common in the Aegean Region.

On a small fishing boat, the thruster moves the boat and the steering is achieved with tiller that is in fact a handle connected to change the direction of the force generated by the thruster. In this small boat example, area available for mounting photovoltaic array on boat is about $1.21 \, m^2$ and five semi-flex mono Silicon PV modules are mounted on a carbon fiber enforced polymer composite plate top. In Fig. 7.9, mounted PVs and gauges for instantaneous measurements can be seen. Each module has 16% efficient 24 monocrystalline PV cells and one cell dimension is 100 mm × 100 mm. Total weight of the modules are 1.5 kg and can generate 192 Wp. PV modules charge two 12 V 26 Ah lead acid type batteries which are used for storing the electric energy for propulsion and for lighting as might be seen in electrical schematic in Fig. 7.10. Batteries are totally 14.5 kg and placed on the front side of the boat in order to balance the load. Circuit breakers and MPPTs are mounted in rear left side of the boat and isolated from water using an IP66 protection class box that is placed inside the boat shell. Electric motor is selected from among outboard electric motors that are suitable for the purpose of the fishing boat. Outboard electric motor is a 540 W brush type DC motor and it can be controlled in five forward speed levels and three speed levels of reverse. The weight of the motor is approximately 24.6 kg. Experimental MPPT have wireless communication and sends date to a laptop computer. In addition to this, two gauges are placed on the board of front side of the boat for battery voltage and motor current. Based on experiments on it, it can provide 5.34 km of range without solar energy and reach to 6 km/h which can be used as a small trolling boat or for touristic purposes.

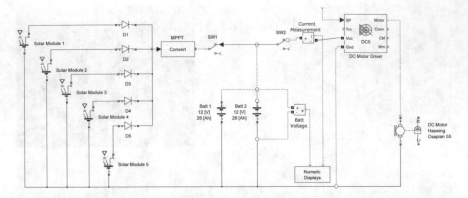

Fig. 7.10 Electrical schematic of the solar angler boat

(a)   (b)

Fig. 7.11 Turanor planet solar boat (Wikipedia, 2022). (**a**) Turanor Planet Solar Boat is a catamaran style boat (**b**) The length of Turanor Planet Solar Boat is 31 meters

There are also bigger solar boats which are used to transportation or for touristic purposes but main purpose of them is to show solar energy harvesting can be implemented on boats or ship. One of the most known giant solar boat is Planet Solar that was developed in 2010. MS Tûranor PlanetSolar is the largest solar-powered boat ever built. Its 500 solar panels can provide 120 kW of energy, allowing the ship to travel around 5 knots cruising speed. A photo of planet solar can be seen in Fig. 7.11.

In last decade, aesthetic, efficient and technological solar boats and planes have been designed and completed their missions to improve that in future can be set up with solar energy solutions. Solar Impulse is one of them. Solar Impulse is a solar plane that has travelled around the world using only energy generated by sun. It has 63.4 m of total wing length and that is exactly the same of an Airbus A340. The solar plane is now in Cité des sciences et de l'industrie in Paris Fig. 7.12.

Mr. Piccard's words about the solar plane Solar Impulse in which he travelled around the whole world is important in this context: "Solar Impulse was not built to carry passengers, but to carry a message."

Fig. 7.12 Solar Impulse (Cité des sciences et de l'industrie, Paris, photo : A. Gören)

7.4 Solar Pumping

Solar power is one of the most cost-effective energy used in pumping systems for irrigation, drinkable water supply and desalination. PV solar pumping systems have increased over the years and become the viable solution due to advantages such as:

- use of free renewable energy.
- no need of big infrastructure.
- less maintenance.
- low running costs.

A solar water pumping system does not necessarily need additional batteries to store energy as the pump can operate during the day to store water into a tank for later use, mainly during the night (Fig. 7.13).

Main applications of solar pumping are:

- Agriculture and forestry irrigation.
- Irrigation for desert plant, urban gardens, parks, greenhouses.
- Fish pond water level maintenance aeration.
- Swimming pool circulation and filtration.
- Surface water pumping for landscaping streams, fountains and waterfalls.
- Supply water for livestock, islands, remote areas.
- Pumping in wastewater treatment, sea water for desalinization.
- Flood water pumping.
- Water pumping for tank storage.

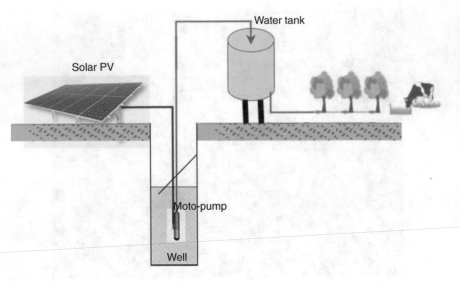

Fig. 7.13 Farm irrigation

7.4.1   Design Aspects

The process of choosing the PV configuration and associated motor-pump is an iterative process based on water rate, maximum total dynamic head and manufacturer's data-sheets which are generally incomplete. Solar PV for irrigation should be associated with water storage for better system performance.

The PV panels take energy from the sunlight and generates DC electricity, which is then directed, through a controller or not, to the pump-motor which makes the water flow from a source through a pipe to a discharge point.

The choice of motor-pump assembly depends on water volume needed, efficiency, price and reliability. DC motors are an attractive option because they are more efficient than AC motors and they don't need AC conversion. On the contrary, AC motors are more available, cost effective and need less maintenance and are generally recommended for power over 2 kW. In both cases, the system requires an MPPT and the pump speed is controlled by its voltage.

Batteries can be incorporated into the PV pumping system but the cost and maintenance will increase considerably. A storage water tank is a more efficient and less expensive alternative to ensuring water availability when sunlight is absent. The capacity of the tank will be determined according to the daily water needs and the required autonomy of the system.

In general, the main factors that affect the selection of a solar powered pump-motor are:

1. Daily water requirement (m³/day): The quantities of water required vary daily, monthly, and seasonally. But maximum water requirement should be considered.

2. Total Dynamic Head (TDH) (meters): This is the effective pressure at which the water pump must operate and it is measured in meters. It has two sub-parameters: total vertical lift and total frictional losses. Further, the total vertical lift is the summation of three parameters: elevation, standing water level, and drawdown.

   - The elevation is the measure of the difference between the ground and the height at which the water is to be discharged.
   - Standing Water Level is the difference between the water level in the well and the surface ground.
   - Drawdown is the measure of the height from which the water level drops down due to pumping the water out.

3. Frictional losses (meters) due to the size of the pipe, type of fittings, air present in the pipe, flow rate, etc.

The system design can be done in five steps as follows;

1. Determine the daily water requirement in m³/day.
2. Calculate H_d, the TDH required for pumping the water.
3. Get the total hydraulic energy required per day (Watt-hour/day) for pumping the water.
4. Estimate the solar radiation available at the site.
5. Size the PV modules required, the motor rating, its efficiency, and losses.

The hydraulic energy E_h (J) is calculated using the basic formula for potential energy

$$E_h = mg H_d.$$

In the context of pumping,

- m (kg) is the mass of the fluid (basically water)
- $g = 9.81$ (m/s²)– acceleration due to gravity
- H_d (m) is the distance the fluid has to travel from its source to the delivery point.

Since 1 Wh = 3600 J, the hydraulic energy in Wh/day is given by

$$E_h = \frac{\rho g V H_d}{3600 \eta_h}.$$

where

- $\rho = m/V$ is the density of the fluid. For water, $\rho = 1000$ kg/m³.
- V (m³/day) is the needed water flow per day.
- η_h is a coefficient < 1 to count for hydraulic losses.

In a standalone solar pumping system, this hydraulic energy E_h is provided by the PV solar installation during the average daily peak sunshine hours (N_S) which depends on the location. The necessary solar power P_S (W_p) is given by:

Table 7.1 Examples of necessary solar power

V (m³/day)	H_d (m)	P_S (W)
12	7	152.6
40	4	290.7
120	40	8720

$$P_S = \frac{E_h}{\eta N_S}.$$

where $\eta = \eta_h \eta_S$, η_S being a coefficient < 1 to count for panel efficiency and power losses.

By considering numerical values for the constant coefficients, P_S can be rewritten:

$$P_S = \frac{2.725 V H_d}{\eta N_S}.$$

7.4.2 Examples

In the following examples, we set $\eta = 30$ and we consider the average number of sunshine hours as $N_S = 5$. With these values, Table 7.1 gives the necessary power for different V and Hd.

As example of market pumping system, a kit including all components (PV, DC-pump, cables...) with

$$V = 5\text{m}^3/\text{day}, \quad H_d = 40\,\text{m}, \quad P_S = 230\,\text{W}$$

costs 1.330€.

The interested reader may find a real life example of solar pumping in [8] and many other applications in the literature.

7.5 Solar Lighting

Solar lighting has been used since a long-time for projects in need of temporary and portable lighting such as in road construction/maintenance, traffic signs and in remote areas without access to the grid. With energy-efficient and cost-effective LEDs, PVs and batteries, solar lighting is increasingly adopted for residential, commercial and industrial lighting of streets, parking lot, pathways, etc. In fact, solar lighting provides many benefits:

- Use of sustainable and Carbon free energy;
- Low cost of maintenance;
- No energy bills;
- Easy fault diagnosis because each light is independent;
- No trenching, no electrical cables between the poles, no grid-tie connection, no substation;
- The technology is well understood and duplicates easily;
- They are versatile and portable if needed;
- The components are widely available and are cost-effective.

The main functions of outdoor lights are to ensure safety, comfort and secure environment during the night. Additional features include aesthetics, highlighting the architectural heritage, providing animation as well as other functionalities such as CCTV, WiFi, 5G. However, excessive lighting can cause wildlife disturbance and discomfort to the neighborhood.

LED lamps are the most advanced technology for lighting due to their advantages:

- Energy efficiency;
- No heating time to switch on;
- Longevity: 50,000 h of runtime then efficiency drops to 70% till 90,000 h of operation;
- Failure resilient if they are connected in parallel to the main power so if one LED fails, the rest is not affected.
- Flexible driveability;
- They allow a wide range of fixtures for more aesthetics. The fixtures are what houses a light and is usually made of several LED fixtures embedded into a full array.

To optimise the lighting system operation, it is possible to adapt it to the needs. This is referred to as smart lighting which yields greater energy savings. In fact, there is no need to spend energy lighting if natural light is sufficient or if nobody or no object is detected in the vicinity. Therefore, the idea is to implement dimming profiles following a timeline and/or use sensors for ambient light and motion/presence detection and then to set the LEDs to a minimum threshold for minimal functions and to its maximum when necessary.

7.5.1 Lighting Characteristics

There are many factors which characterize lighting (Fig. 7.14)

Luminous flux lumen (lm):   It is a measure of the total quantity of visible light emitted by a source per unit of time. Luminous flux differs from power (radiant flux) in that radiant flux includes all electromagnetic waves emitted, while

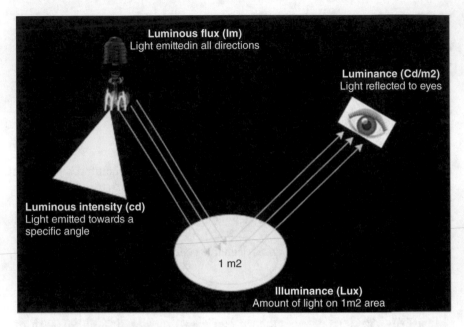

Fig. 7.14 Light characteristics

luminous flux is weighted according to a model (a "luminosity function") of the human eye's sensitivity to various wavelengths.

Illuminance (lux):   This is the quotient of the illuminous flux incident on an element of the surface at a point of surface containing the point, by the area of that element. The illuminance usually qualifies the lighting level and affects the appearance on a specified plane. One lux is one lumen per square metre. Lux can be measured with a luxmeter. Typical values of illuminance are given in Table 7.2.

Luminous intensity (Candella, cd):   It is the measure of light emitted in a specific direction. 1 candela corresponds to the luminous intensity of a candle flame for an observer located at a distance of 1 m.

Luminous efficiency (Lumen/Watt, lm/W):   This is the ratio of luminous flux emitted by a lamp to the power consumed by the lamp. It is a reflection of efficiency of energy conversion from electricity to light form. For instance, an LED produces an average of 70–100 lm/W.

Beam Angle.   It is the angle at which light is distributed from a light source. The specific degree measurement from a lamp varies depending on the manufacturer, but a beam angle falls under four categories: Spot (5–20°), Flood (21–35°), Wide Flood (36–49°), Very Wide Flood (50–120°)

Luminous Intensity Distribution Pattern: This pattern shows the direction and intensity of the light from the light source. Two graphs common in displaying the light patterns are called the plane system and the polar curve.

Table 7.2 Typical values of illuminance

General domestic lighting	120 lux
Classroom	300 lux
Conference and meeting rooms	500 lux
Full moon night	0.5 lux
Well-lit night street	20–70 lux
Streets, roads and highways	15–50 lux
Office	200–400 lux
Outdoors in full sun	50,000–100,000 lux

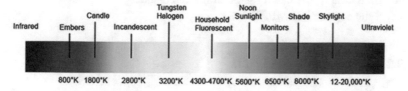

Fig. 7.15 Light color temperature

Color Temperature: The color temperature corresponds to the temperature of the incandescent black body which would give light of equivalent colorimetric composition (Fig. 7.15).

Color rendering index (CRI):   The CRI measures the ability of a light source not to distort the colors with respect to sunlight *i.e.* to reveal the colors of objects in comparison with a natural light source. It is determined by the light source's spectrum. High values of CRI are desirable, the maximum being 100.

7.5.1.1 Design Aspects

Solar lighting systems are effective and have a great potential in many applications. The design is straightforward for the energy part and is made easy for the lighting part using free software such as Dialux https://www.dialux.com and Relux https://relux.com. Components are widely available at a competitive cost, making the investment worthwhile.

The main components in solar smart lighting are:

- Solar panels, MPPT, charger and battery to be sized with respect to the load.
- Driver to control the LEDs.
- Sensors to measure the energy consumption, the illumination level of surrounding and motion detection (LDR, PIR, Ultra-sonic sensors).
- Communication, using DALI protocol PLC (Power Line Communication) technology or wireless protocols such as GPRS, ZigBee, Bluetooth, LORA.
- Monitoring system to perform the required lighting profile and dimming scenario.

The sizing of a solar lighting system follows the following steps:

1. Setting up the required power P_L of the luminaires depending on the type and class inspired by the recommended values in the standards in the considered area (parking, pedestrian, cycling, main road...)
2. Estimation of the daily energy $E_L = P_L \times T$ (Wh/day) actually needed to guarantee that the light will be available during T hours at night.
3. Determination of the daily solar energy E_S available in the considered location.
4. Calculation of the power P_{PV} (W_p) of the PVs to be installed.
5. Determination of the battery technology and capacity.
6. Select the appropriate MPPT.

7.5.1.2   Example 1: Street Lighting

Let us illustrate the calculations for a solar street light with the following configuration:

- Lamp of 60 W power working 7 h per night.
- The location captures 8 sun-hours per day.
- Backup battery: 2 days of operation.

With these figures, the energy needed is $60 \times 7 = 420$ Wh/day. Therefore, to cover 1 day and keep energy stored for 2 more days, we need $3 \times 420 = 1260$ Wh. This energy must be collect during the 8 sun-hours/day by PV powered at $1260/8 = 157.5$ W. Let us select 2 PVs of 80 W_p operating at 24 V. Therefore, the battery capacity is $1260/24 = 52.5$ Ah. We select at least a capacity of 60 Ah.

In practice, we need much more energy because:

- Batteries can't deliver all its stored energy in order to not compromise its own health and lifespan.
- The PV efficiency decreases with high temperature.
- Conversion, MPPT and driver components induce energy losses.

Consequently, PVs and batteries have to be oversized to compensate by higher energy production and storage. However, a dimming profile can be used to optimise energy use and work with limited sizing of components. Namely, the 7 h full lighting could be replaced by a dimming profile as follows:

$$3 \text{ h at } 100\%, 2 \text{ h at } 60\%, 2 \text{ h at } 20\%.$$

Therefore, the energy needed is $60 \times (3 + 0.6 \times 2 + 0.2 \times 2) = 60 \times 4.2 = 252$ Wh/day, 40% less than the 100% full 7 h lighting. We see that dimming can compensate under-sizing and saves energy.

7.5.1.3 Example 2: Outdoor Solar Lamps

Consider the following configuration:

20 W LED-220 V for 10 h of night light and 4 h of sunlight.
Therefore,

- The energy needed is 200 Wh.
- Solar panel power is 200 Wh/4 h = 50 W.
- 12 V lead-acid battery capacity: 200 Wh/12 V = 16.7 Ah. To avoid discharge less than half the battery, the capacity must be doubled. We select a 12 V/34 Ah lead-acid battery.
- Charger: 50 W/12 V = 4.2 A. We select a 12 V/5 A charger.

7.5.1.4 Example 3: Portable Solar Lamps

Portable solar lamps are used in areas where electricity is not available. An example of a self-contained portable solar lamp marketed in rural and remote areas of Africa has the following features:

- average luminous flux of 70 lm.
- battery capacity of 1500 mAh under 3.2 V.
- 1.5 W_p embedded solar cells.

If we assume an average of 5 h sunlight, and LEDs consuming 1 W for 70 lm, then a daily full solar charge is $1500 \times 3.2 = 4.8$ Wh which can ensure 4.8 h of light.

Alternatively, for even cheaper, one can use a marketed solar power bank with a capacity of 20,000 mAh, 18 W foldable solar panels and also include a fast charger and USB ports to supply different lighting systems or charge a phone and so.

Under the same assumptions, the alternative power bank could harvest $18 \times 5 = 90$ Wh solar energy and therefore is equivalent to more than 18 above self-contained portable solar lamps.

References

1. Pieter Gagnon, Robert Margolis, and Caleb Phillips. Rooftop photovoltaic technical potential in the United States. 2019.
2. Deutsche Gesellschaft fur Sonnenenergie (DGS). *Planning and Installing Photovoltaic Systems: a Guide for Installers, Architects and Engineers*. Earthscan, 2006.
3. Firdaus Muhammad-Sukki Nazmi Sellami Samuel Amo Awuku, Amar Bennadji. Myth or gold? the power of aesthetics in the adoption of building integrated photovoltaics (bipvs). *Energy Nexus*, 4, 2021.
4. Enas Taha Sayed Mohammad Ali Abdelkareem Tabbi Wilberforce A.G. Olabi Hussein M. Maghrabie, Khaled Elsaid. Applications and challenges, sustainable energy technologies and assessments. *Building-integrated photovoltaic/thermal (BIPVT) systems*, 45, 2021.

5. Helen Rose Wilson Veronique Delisle Rebecca Yang Lorenzo Olivieri Jesús Polo Johannes Eisenlohr Benjamin Roy Laura Maturi Gaute Otnes Mattia Dallapiccola W.M. Pabasara Upalakshi Wijeratne Nuria Martín-Chivelet, Konstantinos Kapsis. Building-integrated photovoltaic (bipv) products and systems: A review of energy-related behavior, energy and buildings. *Building-integrated photovoltaic/thermal (BIPVT) systems*, 262, 2022.
6. N. Mahdavi Tabatabaei N. Blaabjerg F. Kurt E. (eds) Gören, A. In Bizon. Solar energy harvesting in electro mobility. In *nergy Harvesting and Energy Efficiency*, volume 37. Lecture Notes in Energy. Springer, 2017.
7. Goren A. Ezan M.A. (Korkut, T.B. A cfd study on photovoltaic performance investigation of a solar racing car. 2020.
8. Thulebona Glenacia Gumede, Andre T. Puati Zau, Mokopu Jayluke Malatji, and SP Daniel Chowdhury. Design of stand-alone solar pv system to pump water to the mtubatuba village. In *2019 IEEE AFRICON*, pages 1–6, 2019.

Chapter 8
Feasibility Assessment of Solar Energy Projects

8.1 Feasibility Studies

A feasibility study is a set of investigations that determines whether a certain project satisfies the requirements for implementation and gives recommendations on whether the project should be implemented and under what conditions it should be implemented. Depending on the nature of the project the feasibility assessment may range from a simple and informal process of decision making by a residential owner based on cost and performance estimates provided by the vendor, to a multi-stage and highly comprehensive feasibility analysis process where different aspects and options are carefully considered and formally reported to support decision makers.

The objectives of the feasibility studies include typically:

- give focus to the project;
- provide valuable information for a decision to go ahead or not;
- narrow the alternatives;
- increase the probabilities of contributing to the success by identifying weaknesses at an early stage;
- enhance the success rate by evaluating multiple alternatives;
- consider the life cycle and impact of the project.

Feasibility studies offer the opportunity to make the correct decisions before committing time, money and business resources to an idea that may not work in the way that was originally intended. They help avoid additional costs to correct flaws and remove limitations that could have been identified earlier. They may also help identify new possibilities, opportunities and solutions that might never have otherwise been considered. The key aspects of solar energy feasibility studies are discussed in the following sections, including technical, financial, environmental, legal and social aspects.

A. Rachid et al., *Solar Energy Engineering and Applications*, Power Systems,
https://doi.org/10.1007/978-3-031-20830-0_8

8.2   Technical Aspects

There are a number of considerations relating to the site and the technologies to be used when assessing the feasibility of solar energy projects.

- A performance evaluation of the system to obtain an accurate projection of the solar plant's energy output capacity.
- The best position and orientation for a solar array to maximise system performance, with due consideration of site constraints.
- Surveys for ground-mounted, rooftop based and building/roof integrated systems;
- Options for solar panel mounting, including recommendations for any structural work needed on roofs and buildings due to the additional loads posed by solar panels and mounting structures.
- Roof measurements and condition evaluations.
- In the case of solar PV, cable lengths are assessed, and the integration of a new solar system with existing electrical networks is evaluated. In the case of solar thermal systems, pipe length estimates and an assessment of the integration of the solar thermal system into existing heat networks are necessary.
- Assessment of the capacity and condition of existing power supply cables, and electrical compatibility checks, in the case of solar PV systems; or capacity and condition of existing heat network piping in the case of solar thermal systems.
- Local shading considerations in relation to nearby structures or objects that might affect the performance of a solar PV installation.
- Assess the need for any electrical, roof, ground or any other works that may be required prior to the installation of the solar energy system.
- Examine any potential hazards or risks associated with the solar system and its installation.
- Identify any site access restrictions.
- Identify any potential grid connection restrictions and the approval process by the utility company.
- Assess any planning permission considerations and identify the details of any applicable building regulations and the authority responsible for them (also see social and legal aspects).

Solar modules are either mounted on fixed-angle frames or on sun tracking frames. Fixed frames are simpler to install, cheaper and require less maintenance. However, tracking systems can increase yield by over 40%.

The solar resource expected to be available for the lifetime of a solar plant can be determined by analysing historical solar resource data for the site. The plane of array irradiance plays a role in obtaining a preliminary approximation of a plant's power or heat output. As a result, the accuracy of any solar energy yield projection is strongly reliant on the solar resource dataset that is employed. Because project profits are dependent on the energy yield of the solar installation, using trustworthy historical resource data is a critical step in the development process and is required for project finance.

Satellite-derived data and land-based measurements are the two main sources of solar resource datasets. Because both sources have advantages and limitations, the decision as to which one to use is normally based on the individual site as well as the organization's preferences and procedures. Because solar energy is variable, an understanding of inter-annual fluctuation is essential. To compute the variation with an appropriate degree of confidence, 10 years or more of data is usually required.

In the northern hemisphere, a surface slanted at an inclination towards the south receives more total yearly global irradiation than a flat surface. This is due to the fact that a surface slanted to the south faces the sun more directly for a longer amount of time. The opposite is true in the case of a location in the southern hemisphere. The amount of irradiation received can be calculated using simulation software for every tilt degree. The best tilt angle is determined by latitude, but it can also be influenced by local weather patterns and plant layout combinations. The irradiance reflected from the ground towards the modules, which is dependent on the ground reflectance, is generally factored into this estimate. Information about the sun resource and temperature conditions of the site, as well as the layout and technical specifications of the plant components, are required to accurately estimate the energy produced by a solar energy system. To model the interaction of temperature, irradiance, shading, and wind cooling on the panels, specialised software packages are frequently utilised.

8.2.1 Modelling Case Study

The Eco-House, shown in Fig. 8.1, is a research facility at the University of Portsmouth, UK, consisting of an instrumented 3 bedroom household that is used for research in energy efficiency and building performance. With funding from the Interreg 2 Seas SOLARISE project, the Eco-House was provided in 2021 with a 5 kW solar array, a 13.5 kWh smart battery storage system, energy monitoring and other technologies for research and demonstration purposes. Table 8.1 describes the key items of equipment installed at the Eco-House. The PV array has an azymuth orientation of 20° and each module is tilted by 15°.

A computer-based model of the system at the Eco-House has been implemented using the software package PVSYST, including representations the specific JA Solar solar modules and their layout, SolarEdge power optimisers, SolarEdge inverter, Tesla Powerwall 2 battery, and the structures around the solar array that cause shading. The demand of the house was represented with constant power resulting in a total daily energy demand of 10 kWh. A perspective of the solar array and its surroundings is shown in Fig. 8.2. The location of the house is 50.7977° North, 1.0978° West, and the altitude is assumed to be 12 m above sea level. The model considers ambient temperature dependencies, and the losses that occur at different parts of the system, including those associated with shading. The model used the

Fig. 8.1 This image shows the installed 5 kW photovoltaic solar array mounted on the adjacent lawn by the Eco-House, which is shown to the right of the solar array

Table 8.1 Key equipment installed at the Eco-House

Item description	Quantity
JA Solar HAN60S10 335 W monocrystalline solar module	15
Renusol ConSole CS+ standalone ballasted PV-mounting system	15
SolarEdge P370 power optimizer	15
SolarEdge SE5000H single phase inverter	1
Tesla Powerwall 2 battery, single phase 230 V AC connection, 13.5 kWh usable energy capacity	1
Tesla Powerwall 2 backup gateway	1

Meteonorm v. 8.0 database for Portsea, using synthetic data for hourly values from the measured monthly quantities (Portsea MN80 SYN). The simulation assumes that the battery charge and discharge are performed to maximise the self-consumption of solar energy, which is a commonly used setting.

Table 8.2 shows various energy quantities predicted by the model over one generic year, divided into individual months. The energy yield of the solar array is estimated to be 3952.6 kWh over the first year. After loses, the available energy on the AC side of the inverter is 3897 kWh over the first year, of which 2696.7 kWh (69.2%) are self-consumed at the house, 833.5 kWh (21.4%) are exported to the grid, and 366.6 kWh (9.4%) are lost at the battery. The photovoltaic system provides 74% on the annual electricity consumption of the house over a year, which is 3644 kWh. Figure 8.3 shows a loss diagram illustrating the energy flow through the system and

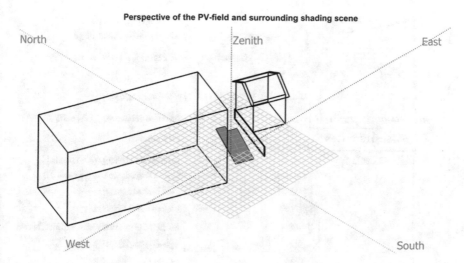

Perspective of the PV-field and surrounding shading scene

Fig. 8.2 Perspective of the photovoltaic array at the Eco-House and surrounding structures, including the Eco-House itself, the wall to the right of the array and the Dennis Sciama building to the left of the array

Table 8.2 Technical quantities predicted by the model over the first year, divided into individual months

Balances and main results

	GlobHor	DiffHor	T_Amb	GlobInc	GlobEff	EArray	E_User	E_Solar	E_Grid	EFrGrid
	kWh/m²	kWh/m²	°C	kWh/m²	kWh/m²	kWh	kWh	kWh	kWh	kWh
January	25.3	15.18	5.23	35.7	27.4	129.8	309.5	118.6	0.0	190.9
February	39.6	28.57	5.21	47.3	35.5	170.1	279.6	147.5	0.0	132.1
March	84.0	41.96	7.41	99.9	74.9	348.7	309.5	263.8	35.0	45.7
April	133.7	57.18	10.17	148.5	103.1	469.4	299.5	282.9	145.5	16.6
May	163.6	76.12	13.51	170.7	111.9	505.7	309.5	305.4	145.1	4.1
June	179.7	82.74	16.29	183.2	115.9	516.3	299.5	299.5	167.8	0.0
July	173.8	73.39	18.28	177.5	114.9	504.2	309.5	304.7	150.6	4.8
August	136.0	65.71	18.02	145.7	100.9	449.1	309.5	292.5	118.7	17.0
September	99.3	43.83	15.26	115.4	82.4	365.7	299.5	259.7	57.9	39.8
October	56.5	26.11	12.51	71.5	53.7	241.0	309.5	202.0	12.9	107.5
November	30.2	17.50	8.50	41.8	32.0	149.6	299.5	130.1	0.0	169.4
December	19.9	12.99	5.76	28.4	21.7	103.0	309.5	89.9	0.0	219.6
Year	1141.5	541.29	11.38	1265.7	874.2	3952.6	3644.2	2696.5	833.5	947.6

Legends

GlobHor	Global horizontal irradiation	EArray	Effective energy at the output of the array
DiffHor	Horizontal diffuse irradiation	E_User	Energy supplied to the user
T_Amb	Ambient Temperature	E_Solar	Energy from the sun
GlobInc	Global incident in coll. plane	E_Grid	Energy injected into grid
GlobEff	Effective Global, corr. for IAM and shadings	EFrGrid	Energy from the grid

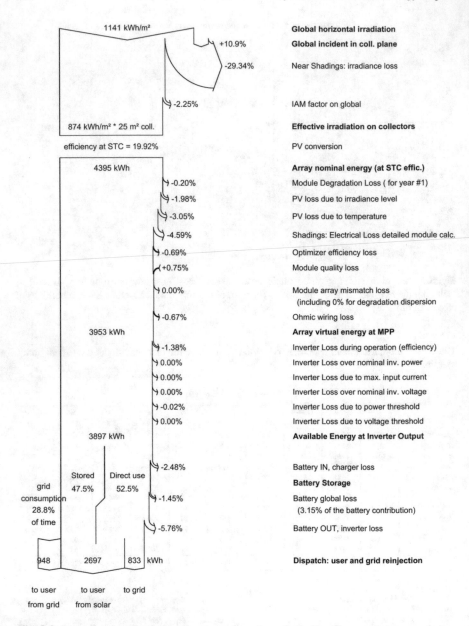

Fig. 8.3 Loss diagram illustrating the energy flow through the Eco-House solar array + battery storage system and associated losses

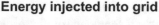

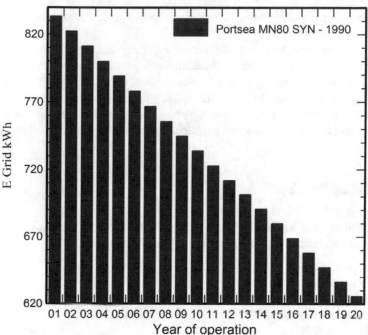

Fig. 8.4 Energy injected into the grid by the Eco-House PV system as a function of the year of operation over 25 years, illustrating the effect of PV module degradation that is represented by the model

associated losses. Figure 8.4 shows the energy injected into the grid by the Eco-House PV system as a function of the year of operation over 20 years, illustrating the effect of the average PV module degradation of 0.4% per year that is represented by the model

8.3   Financial Aspects

The financing of a large scale solar energy project is possible when the solar plant is highly likely to generate enough revenue to pay for debt obligations and all costs of operation and maintenance, and to generate an adequate return for the equity invested [1].

In case of commercial organisations, the decision to proceed with the development of a solar energy project depends mainly on the commercial viability of the project, as determined through a suitable financial analysis. The requirements for return on investment are lower for public organisations, such as local councils, than for commercial organisations. Moreover, in some cases, a negative return on investment can be acceptable if the principal aim of the project is different from generating profits (e.g. in the case of demonstration projects) and the organisation is willing to bear the cost.

Financial analysis will normally consider the expected costs, including investment requirements, operations and maintenance costs, along with expected revenues. The predicted energy yield of the solar plant, which is normally estimated by using engineering calculations, is used as a key figure to estimate revenues. For the sake of brevity, the examples below refer to solar PV systems, but all the financial concepts and measures mentioned here apply also to solar thermal systems.

8.3.1   Capital Costs

The capital costs of a typical solar PV power plant include the following, where the number in parenthesis indicates the proportion of the total costs:

1. Solar panels (35%)
2. Inverters (6%)
3. Mounting structures (10%)
4. Balance of systems (24%)
5. Other costs (25%)

"Balance of systems" includes, for example, cabling, junction boxes, transformers, and switch gear. Other costs include installation, land registry and planning, grid connection, engineering and technical analysis, financial and legal services. Additional costs could also include the cost of land if it is going to be purchased, as well as other infrastructure, such as roads and drainage.

The average costs of a 0–4 kW installation in the UK between April 2020 and March 2021 was GBP 1628 per installed kW, while the mean value goes down to GBP 1088 for larger 10–50 kW installations [2]. Due to the economies of scale of large ground-mounted plants their cost of installation is significantly less than rooftop systems. The capital cost for a typical 10 MW solar plant is in the region of GBP 800,000 per MW installed [3]. Prices of solar installations for all sizes have consistently been decreasing for several years.

Grid connection expenses can be a considerable part of capital expenditure (CAPEX) for larger projects, to the point that they can make the project financially unviable in some situations, although they can vary significantly from project to project. As the grid's capacity to handle extra generation in some locations becomes limited, the impact of grid connection concerns on solar farm project decision-making has grown. Additional costs are incurred when changes to the substation are required to absorb the incoming energy from the solar plant. Grid connection costs are primarily dependent on the point of connection itself.

For large solar farms, grid connection costs can range from GBP 100,000 per MW to GBP 1,700,000 per MW. For smaller solar installations, grid connections are usually below GBP 400 per kW [4]. It is worth noting that smaller projects (typically <50 kW) can often be connected to localised distribution boards (such as in commercial or public buildings) instead of being connected to a substation, which therefore incurs no grid connection cost.

8.3.2 Operation and Maintenance Costs

Because of their simple engineering and relatively modest maintenance required, solar energy projects have lower operation and maintenance costs than other renewable energy and conventional technologies. In the UK, for example, typical operations and maintenance costs are currently approximately GBP 3200/MW per year. Insurance, administration, professional fees, and land rental are among other operational expenses, in addition to labour. During the course of the project, it may be essential to replace equipment such as batteries, inverters, or even panels, and this may be factored in the financial analysis by considering the expected lifetime of each item of equipment.

Typically, solar farms are constructed in rural areas where the land is not very expensive. In the UK, solar developers pay the land owner a rent of the order of GBP 1000 per acre per annum. In the UK, the land rental cost can be estimated to be GBP 10,000 per MW per annum [3].

8.3.3 Revenues and Electricity Tariffs

The predicted annual energy yield directly influences the revenue line in the cash flow estimate. Therefore, accurate energy yield forecasts are crucial when it comes to large-scale projects. An impartial and appropriately qualified solar energy expert must determine the annual energy yield if funding is necessary. The uncertainty of the predicted energy yield is particularly critical, as the annual energy yield significantly impacts the project's viability.

The key revenue stream for most solar power plants is the tariff paid for each kWh of electricity generated. Sometimes there are other sources of revenue, such as renewable energy credits, tax credits, and other financial incentives available to developers. The permanency of such incentives should be assessed carefully, as they are often modified or eliminated by the government. Tariffs and incentives vary from country to country.

8.3.4 Debt Servicing, Capital Repayment and Taxes

Solar energy projects are frequently financed with a mix of debt and equity. Large solar projects often have a 30/70 or similar equity/loan ratio. Loans demand regular payments, which are determined by the interest rate and the quantity of money borrowed at the outset. Therefore, the lender will want to see good cash flow projections that show the project will generate enough money to cover all ongoing costs and interest and capital repayments.

Because the solar system will generate an income and possibly profits, any taxes imposed will have to be compensated by those earnings over the project's lifetime. Local and national legislation and the legal entity that owns and runs the solar plant determine the taxes that must be paid.

8.3.5 Financial Modelling

Examining the project's viability requires an appropriate financial model. Such a model is critical in preparing a project for finance and is an integral part of decision-making. The financial model calculates the essential parameters required for the project's financial evaluation. Furthermore, the model should demonstrate that the project can pay any debts associated with it. The definitions of some of the key financial factors that can be essential in solar energy projects are discussed below.

8.3.5.1 Net Present Value (NPV)

The Net Present Value (NPV) is the difference between the present value of future net cash flows generated by an investment and its initial cost. It can be expressed as follows:

$$\text{NPV} = -C_0 + \sum_{i=1}^{N} \frac{C_i}{\left(1 + \frac{r}{100}\right)^i}$$

where C_0 is the initial cost, C_i is the net cash inflow in period i, r is the discount or interest rate over one period (in percentage form), and N is the number of periods. One period can be 1 year, 1 month, etc, depending on the case. The net cash inflows consider all income streams for each future period, as well as all expenses for each future period. An investment is worthwhile if it creates value for its owners, and an investment creates value if it worth more than it costs considering the time value of money. This implies that the NPV is normally expected to be positive, unless a positive NPV is not a priority given other considerations (for example, when the principal aim of the project is not to generate a profit).

8.3.5.2 Payback Period (PP)

The payback period is the amount of time that it takes for a project to recover its initial cost out of the cash income that it produces. Assuming that time is measured in years and that the annual net cash inflow C_i is the same each year over the duration of the project, the following formula can be used to compute the payback period:

$$PP = \frac{C_0}{C_i} = \frac{\text{Initial cost}}{\text{Annual net cash inflow}}$$

If the expected cash flows is uneven (different cash flows in different periods), the payback formula cannot be used to compute payback period of the project. In such cases, the payback period can easily be computed by tracking the unrecovered investment year by year. Acceptable values of the payback period depend on organisational factors. Payback periods are often within the range of 5–20 years.

8.3.5.3 Return on Investment

The return on investment is defined as the ratio of net benefit against the initial investment, which measures system profitability. A negative ROI indicates that the investment is not profitable. The formula to calculate the ROI is as follows:

$$ROI = \frac{\text{Net benefit at the end of lifetime}}{\text{Total investment}}$$

8.3.5.4 Levelised Cost of Electricity (LCOE)

The Levelised Cost of Electricity (LCOE) is the net present value of the unit-cost of electricity over the lifetime of an electricity generating asset or project. This measure is often taken as a proxy for the average price that the generating asset must receive to break even over its lifetime. This figure is an economic assessment of the cost competitiveness of an electricity-generating system that incorporates all costs over its lifetime, including the initial investment, operations and maintenance, the cost of fuel and the cost of capital.

A net present value calculation is performed such that for the value of the LCOE chosen, the project's net present value is equal to zero. This means that the LCOE is the minimum sales electricity price for an the project to break even. Typically, LCOEs are calculated over 20–40 year lifetimes, and are provided in currency per kilowatt-hour, or per megawatt-hour. See [5] for an online calculator for the LCOE.

There is a specialised variant of LCOE for cooling systems, which is known as the Levelised Cost of Cooling [6]. Moreover, a specialised variant of LCOE has been defined for heating systems, known as the Levelised Cost of Heat [7].

Table 8.3 Financial analysis associated with Eco-House example

	Financial analysis		
Simulation period			
Project lifetime	20 years	Start year	2023
Income variation over time			
Inflation		3.50 %/year	
Production variation (aging)		Aging tool results	
Discount rate		2.20 %/year	
Income dependent expenses			
Income tax rate		0.00 %/year	
Other income tax		0.00 %/year	
Dividends		0.00 %/year	
Financing			
Loan - Redeemable with fixed annuity - 10 years		17602.00 GBP	Interest rate: 3.00%/year
Electricity sale			
Feed-in tariff	Peak tariff	0.0500 GBP/kWh	
	Off-peak tariff	0.0500 GBP/kWh	00:00-06:00
Duration of tariff warranty		25 years	
Annual connection tax		0.00 GBP/kWh	
Annual tariff variation		0.0 %/year	
Feed-in tariff decrease after warranty		50.00 %	
Self-consumption			
Consumption tariff	Peak tariff	0.4300 GBP/kWh	
	Off-peak tariff	0.3000 GBP/kWh	00:00-06:00
Tariff evolution		+10.0 %/year	
Return on investment			
Payback period		13.7 years	
Net present value (NPV)		9395.17 GBP	
Return on investment (ROI)		53.4 %	

8.3.5.5   Financial Modelling Example

As an example, Table 8.3 shows the assumptions and results of the financial analysis of the photovoltaic + battery storage system at the Eco-House, University of Portsmouth, which is described in Sect. 8.2.1. The total initial cost of the installation is assumed to be GBP 17,602, reflecting current market prices. For the purposes of this example, the project is supposed to be financed through a 10 year loan for GBP 17,602 at 3% interest. A constant inflation rate of 3.5% per year is considered. The project lifetime is assumed to be 20 years and a discount rate of 2.2% is employed. The electricity savings are estimated using a tariff of 0.43 GBP/kWh at peak times (06:00–23:59) and 0.30 GBP/kWh at off-peak times (00:00–05:59). The electricity tariff is assumed to increase by 10% each year. A constant feed-in tariff of 0.050 GBP/kWh is considered. An annual maintenance allowance with an initial value of GBP 50.0 is included. The PVSYST software aging tool was employed considering an average PV module degradation of 0.4%/year. No equipment replacement was considered during the project's lifetime. Table 8.4 shows the detailed financial values over the lifetime of the project. Notice that the payback period is predicted to be 13.7 years, the NPV of the investment is estimated to be GBP 9,397.63, the return on investment is 53.4%, and the levelised cost of electricity is estimated at 0.42 GBP/kWh.

Table 8.4 Detailed financial results for the Eco-House example

	Electricity sale	Loan principal	Loan interest	Run. costs	Deprec. allow.	Taxable income	Taxes	After-tax profit	Self-cons. saving	Cumul. profit	% amorti.
2023	42	1535	528	50	0	0	0	-2072	1090	-961	3.3%
2024	41	1581	482	52	0	0	0	-2074	1182	-1815	7.4%
2025	41	1629	435	54	0	0	0	-2076	1273	-2568	12.4%
2026	40	1678	386	55	0	0	0	-2079	1359	-3227	18.2%
2027	39	1728	335	57	0	0	0	-2081	1444	-3799	24.7%
2028	39	1780	284	59	0	0	0	-2084	1525	-4289	32.1%
2029	38	1833	230	61	0	0	0	-2087	1603	-4704	40.1%
2030	38	1888	175	64	0	0	0	-2089	1679	-5049	48.9%
2031	37	1945	118	66	0	0	0	-2092	1752	-5328	58.3%
2032	37	2003	60	68	0	0	0	-2095	1822	-5547	68.5%
2033	36	0	0	71	0	0	0	-34	1889	-4087	76.8%
2034	36	0	0	73	0	0	0	-37	1954	-2611	85.2%
2035	35	0	0	76	0	0	0	-40	2017	-1121	93.6%
2036	35	0	0	78	0	0	0	-44	2077	378	102.1%
2037	34	0	0	81	0	0	0	-47	2133	1883	110.7%
2038	33	0	0	84	0	0	0	-50	2186	3390	119.3%
2039	33	0	0	87	0	0	0	-54	2236	4898	127.8%
2040	32	0	0	90	0	0	0	-57	2284	6403	136.4%
2041	32	0	0	93	0	0	0	-61	2329	7903	144.9%
2042	31	0	0	96	0	0	0	-65	2371	9395	153.4%
Total	729	17602	3033	1414	0	0	0	-21320	36207	9395	153.4%

8.4 Environmental Aspects

The potential environmental impacts of a solar energy plant, which include land use and habitat loss, water use, hazardous material use in manufacturing, landscape and visual impacts, and global warming emissions, can vary greatly depending on the technology used, the location, the scale, and other aspects of the project.

8.4.1 Habitat Loss

When a habitat is destroyed, the plants, animals, and other species that inhabited it have a lower carrying capacity, resulting in population decrease and the possibility of extinction. The process of habitat loss is perhaps the greatest hazard to organisms and biodiversity.

Larger utility-scale solar projects, depending on their location, might cause concerns about land degradation and habitat loss among nearby animals, plants, and other creatures, as illustrated in Fig. 8.5. The total amount of land required is determined by technology, topography, and the intensity of the solar resource available at the location. For instance, in the United States utility-scale fixed PV systems require about 3 hectares per MW of peak capacity [8]. The impact of utility-scale solar systems on habitat loss can be reduced by putting them in low-quality areas, such as old landfills. Smaller solar PV arrays, which can be installed on residences or commercial buildings, have little or no influence on habitats.

Fig. 8.5 Illustration of the notion of habitat loss, where a section of a forest has been replaced by a solar plant. Source: 'Long Island Solar Farm' by Brookhaven National Laboratory is licensed under CC BY-NC-ND 2.0

8.4.2 Ground Concurrency

A solar system should not be used to dispense with potentially useful agricultural land. Although governmental constraints in some countries limit agricultural land use, low-impact solar development and co-location of solar and agriculture provide an alternative to such restrictions. This strategy is currently being studied and has the potential to reduce agriculture displacement by allowing solar arrays, flora, and cattle to coexist on the same piece of ground. Figure 8.6 illustrates the idea of ground concurrency.

Low-impact solar development can be classified into three types to satisfy varied project aims. The first is solar-centric design, which maximises solar output while maintaining low-lying vegetation for ground cover and wildlife habitat. The second type is vegetation-centric design, which is optimised for plant growth while allowing solar arrays to be placed in regions where vegetation would not be harmed. Co-location design is the third category, in which solar and vegetation layouts are built together for maximum dual output. Leaving existing vegetation in place or replacing it with low-growing native vegetation, planning the solar project around natural land contours, and having plant supporting habitats are all examples of these three categories [8].

Fig. 8.6 Illustration of ground concurrency. Source: 'Sheep and solar panels' by Oregon State University is licenced under CC BY-SA 2.0

8.4.3 Water Use

Solar PV cells generate power without the usage of water. Water is, however, needed in the production of solar PV components. Water is used to cool concentrated solar thermal plants (CSP). The amount of water used is determined by the plant's design, location, and cooling system. CSP systems with cooling towers that use wet-recirculating technology require between 600 and 650 gallons of water per MWh of power generated. Water use should be compared to that of local communities, and any influence on local water supplies should be recognised and managed.

8.4.4 Life Cycle Environmental Impact of Solar Energy Systems

In comparison to traditional energy sources, solar energy systems provide considerable environmental benefits, contributing to the long-term development of human activities. However, their widespread use can also have harmful environmental consequences [9].

There are a variety of techniques to assess environmental impact, but one of the most straightforward is to look at the global warming emissions related with the solar installation. While there are no greenhouse gas emissions related with solar energy generation, there are emissions associated with other stages of the solar plant's life cycle, such as production, materials transportation, installation, maintenance, and decommissioning and dismantling. For photovoltaic systems, life-cycle emissions are estimated to be between 32 and 81 g of carbon dioxide equivalent per kWh. Concentrated solar power is estimated to emit 36–90 g of carbon dioxide equivalent per kWh.

Various hazardous compounds are employed in the PV cell production process, the majority of which are used to clean and purify the semiconductor surface. Hydrochloric acid, sulfuric acid, nitric acid, hydrogen fluoride, and acetone are among the chemicals utilised in the semiconductor business. The amount and type of chemicals used are determined by the cell type, the amount of cleaning required, and the silicon wafer size.

A 2012 study evaluated the environmental and energy impacts of battery production. The research focused on five types of battery: lead-acid, lithium-ion, nickel-cadmium, nickel-metal-hydride and sodium-sulphur [10]. Lithium-ion batteries had the largest impact on metal depletion, specifically of lithium which is used in the battery itself. It is estimated that per MJ of capacity, lithium-ion batteries are about half as toxic to humans as lead-acid batteries, and less toxic than nickel-cadmium batteries. Nickel-metal hydride and sodium-sulphur batteries are the least toxic to humans. The results indicate that the production of lithium-ion and nickel-metal-hydride batteries use the most energy and produce the most greenhouse gas emissions.

A possible source of information on environmental impact of solar energy systems is the Ecoinvent database [11], which provides life cycle inventory information for many products and processes, including many common components of solar energy systems. This database has been used in a number of studies on the assessment of the environmental impact of solar energy systems [12].

8.4.5 Landscape and Visual Impacts

The visibility of the solar panels within the larger landscape, as well as accompanying implications on landscape designations, character types, and adjacent communities, are examples of landscape and visual impacts. Consideration of layout, size, and scale during the design process, as well as landscaping and planting to screen the modules, are all common mitigation measures to reduce impacts. The possibility of glint and glare should also be taken into account as part of the environmental impact assessment.

8.4.6 Reduction in Carbon Emissions

The reduction in carbon emissions because of the substitution of fossil fuel generation that a solar installation enables has a positive impact in the environment. A simple initial calculation of carbon savings from solar photovoltaic installations involves the assumption that all solar electricity directly replaces electricity produced by large power stations. A common way of calculating this is by using the 'average grid carbon intensity', which is the average amount of CO_2 emitted for each kWh of electricity produced for the power grid. For example, the UK average grid carbon intensity is estimated at 210.95 gCO_2/kWh for 2021.[1] Please note that this average figure may vary from country to country and from year to year. A more conservative approach assumes that solar power replaces electricity produced by efficient gas power plants commonly used as rapid response supply to ensure grid balance. For example, it is estimated that these efficient gas plants in the UK currently emit 394 gCO_2/kWh.

As an example, Fig. 8.7 shows the CO_2 emission balance of the photovoltaic system at the Port-Eco-House, University of Portsmouth, which is described in Sect. 8.2.1. This emissions balance was calculated with the software package PVSYST. Considering the original negative saved emissions associated with the

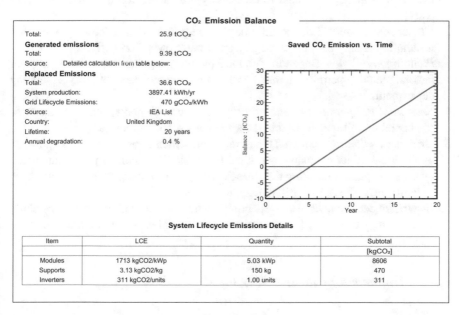

Fig. 8.7 This figure shows the CO2 emission balance of the photovoltaic system at the Eco-House, University of Portsmouth. Notice that the system is predicted to become carbon neutral after approximately 5 years

---

[1] https://data.nationalgrideso.com/carbon-intensity1/historic-generation-mix.

components (PV modules, supports, inverter), and the accumulated saved emissions over time due to the solar generation, notice that the system is predicted to become carbon neutral after approximately 5 years, and to produce net CO_2 savings from then on. Although the emissions associated with the initial installation are 9.39 tons of CO_2, over a period of 20 years, the system is estimated to produce a net CO_2 savings of 25.9 tons of CO_2, by replacing an estimated of 36.6 tons of CO_2 through solar PV generation.

8.5   Social and Legal Aspects

8.5.1   Legal Aspects

Permits and licensing for solar energy installations can be a lengthy process involving multiple agencies in the central and local governments. Depending on the characteristics of the project and the local or national legislation some (or all) of the following may be needed:

1. *Land lease agreement.* This agreement may be required for ground mounted installations in cases the land to be used by the installation is owned by a third party;
2. *Site access permit.* This permit may be required when accessing the solar installation is only possible through a third-party owned property.
3. *Planning permission.* Depending on local planning regulations, and various other factors, planning permission may be required from the local authority's planning department.
4. *Environmental permit.* Depending on local environmental regulations, an environmental permit may be required for the construction of a solar plant.
5. *Grid connection agreement.* This agreement, which is processed and approved by the local utility company, is critical to enable the export of power generated by the solar plant. In making the site selection, the proximity to the grid will have a significant influence on grid connection costs.
6. *Electricity generator license.* Depending on the location and the size of the plant, a license may be required to generate electricity and export it into the grid.

8.5.2   Impacts on Cultural Heritage

Effects on the setting of designated heritage sites or direct impacts on underground archaeological deposits resulting from earth movements during the construction phase of a solar project are examples of potential cultural heritage impacts. Field surveys should be conducted prior to construction to determine the heritage and archaeological components at or near the site if they have been previously identified as a potential issue. Examples of possible mitigation methods include careful site layout and design to avoid areas of cultural heritage or archaeological importance and the execution of a procedure that addresses and safeguards any unexpected cultural heritage finds made during the construction period.

If a person or organisation is proposing to carry out work on a listed or protected building, they must adhere to the standards outlined in relevant national or regional planning regulations, and planning permission is most likely required. When assessing the influence of a proposed development on the significance of a designated heritage building, the asset's conservation should be prioritised. For example, many historic structures have steep roof slopes that could be used to generate electricity using solar panels or slates. Such roofs are frequently highly visible and contribute to the building's character. Solar panels may be more easily accommodated where the building has shallow-pitched roofs that are mostly hidden from view, or internal roof slopes that are not visible from ground level.

8.5.3 Community Involvement

Community involvement is an essential aspect of large-scale solar energy projects. It should be a continuous process that includes disclosing pertinent project-related information to communities that might be affected by the project. Community involvement aims to establish and maintain a positive and productive connection with the communities surrounding the project site over time and to identify and minimise any potential adverse effects on those communities.

8.5.4 Energy Independence

When an organisation, community or household is able to produce enough of its own energy to meet its own demands, then it is referred to as being energy independent. Energy independence can be desirable as it isolates the organisation, community or household from price fluctuations, quality of supply issues, and it can bring clear financial advantages. Even if the organisation, community or household is not fully energy independent, just increasing the degree of independence from external energy suppliers can be a desirable consequence of a solar energy project.

References

1. World Bank Group International Finance Corporation. Utility-scale Solar Photovoltaic Power Plants, 2015.
2. UK Government. Solar PV Cost Data, 2021.
3. Solarmango. What is the installation cost of utility scale solar power plant ($/mw) in the uk?, 2015.
4. UK Department of Energy and Climate Change. DECC small scale generation costs update, 2015.
5. National Renewable Energy Laboratory. Simple Levelized Cost of Energy (LCOE) Calculator Documentation., 2008.
6. R. Gabbrielli. Performance and economic comparison of solar cooling configurations. *Energy Procedia*, 91:759–766, 2016.
7. O. Gudmundsson. Cost analysis of district heating compared to its competing technologies. *WIT Transactions on Ecology and The Environment*, 176, 2013.
8. National Renewable Energy Laboratory. Solar sheep and voltaic veggies: Uniting solar power and agriculture.
9. N.F. Tsoutsos. Environmental impacts from the solar energy technologies. *Energy Policy*, 33:289–296, 2005.
10. M. McManus. Environmental consequences of the use of batteries in low carbon systems: the impact of battery production. *Applied Energy*, 93:228–295, 2012.
11. Ecoinvent. Ecoinvent Database, 2022.
12. Alexis de Laborderie, Clément Puech, Nadine Adra, Isabelle Blanc, Didier Beloin-SaintPierre, Pierryves Padey, Jérôme Payet, Marion Sie, and Philippe Jacquin. Environmental impacts of solar thermal systems with life cycle assessment. volume 57, pages 3678–3685.

Chapter 9
Solar Thermal Energy Systems

9.1 Introduction

There are several applications that use heat from solar energy. This chapter has therefore been structured by presenting the main applications that use solar thermal energy. After some generalities about solar thermal energy systems, water/air heating application and power generation application have been presented.

Basically, solar thermal energy systems transform solar radiation into heat to be used for its intended application. The main element of any solar thermal system is the collector. It absorbs the solar energy, transforms it into thermal energy, and transfers the thermal energy to a heat transfer fluid (such as water, oil or air). The collected energy can be used for water heating, air conditioning, electricity generation through heat exchanger or storage during day so that it can used in evening/night. A layout has been shown in Fig. 9.1.

There are different ways to distinguish solar collectors. It can be on the type of heat transfer fluid or whether they are covered or uncovered. But, the most common way is to distinguish them into tracking or non-tracking. Tracking collectors mainly consist of concentrators which concentrate the sun rays from a larger area to a smaller receiver area which makes them suitable for high-temperature applications. Non-tracking collectors can be low concentrating or flat plate which are suitable for low-temperature applications. To go a bit further in this categorisation, here is an exhaustive classification of the existing solar collectors shown in Fig. 9.2.

As previously said, they are separated from non-tracking to tracking ones. Then, within the tracking solar collectors, there has been distinction for those who follow the Sun on 1-axis of tracking from those on 2-axis. But, before describing most of them, and giving precisions on the purpose they are used for, some technical details for each of the collectors have been presented in Table 9.1, such as the type of absorber, the concentration ratio for each technology i.e. the ratio of the aperture area over the absorber area and an indication of their temperature range. Globally, one can see that the solar collectors cover a large range of temperature.

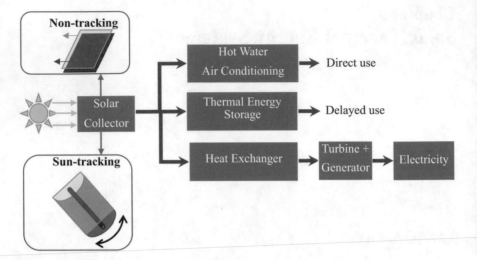

Fig. 9.1 Layout for the working of the solar thermal systems

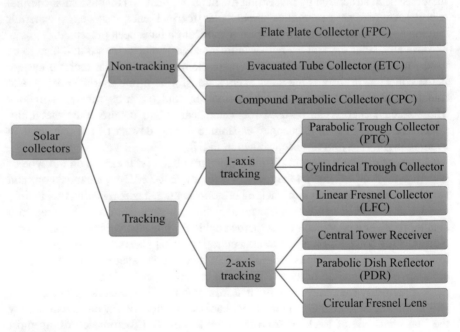

Fig. 9.2 Classification of solar thermal collectors

Table 9.1 Solar thermal collectors [1]

Motion	Collector type	Absorber type	Concentration ratio[a]	Indicative temperature range (°C)
Stationary	Flat-plate collector	Flat	1	30–80
	Evacuated tube collector	Flat	1	50–200
	Compound parabolic collector	Tubular	1–5	60–240
1-axis tracking	Compound parabolic collector	Tubular	5–15	60–300
	Linear Fresnel reflector	Tubular	10–40	60–250
	Cylindrical trough collector	Tubular	15–50	60–300
	Parabolic trough collector	Tubular	10–85	60–400
2-axis tracking	Paraboloid dish reflector	Point	600–2000	100–1500
	Heliostat field collector	Point	300–1500	150–2000

[a]Concentration ratio is defined as the aperture area divided by the receiver/absorber area of the collector

More efficient collectors are the tracking ones. It can also be seen that higher the concentration ratio of the optical device, higher the level of temperature reached.

9.2 Application of Solar Thermal Energy for Water/Air Heating

The most common application of the solar thermal systems is the water/air heating. For this application, non-tracking solar collectors are mainly used. They are stationary and thus, permanently fixed in position and do not track the sun. Three main types of collectors fall into this category are as follows

– Flat-plate collector
– Evacuated tube collector
– Compound parabolic collector

Fig. 9.3 Schematic of the flat plate collector

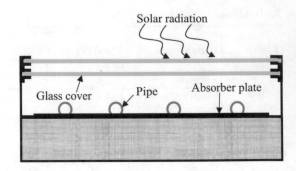

9.2.1   Flat Plate Collector

A schematic of the flat plate collector has been shown in Fig. 9.3. In this collector, the solar radiation transmits through the glass cover and absorbed by the absorber plate which is black in colour to enhance the absorption. The metal pipes are attached to the plate. Water or air can be run through the pipes to collect the heat and to produce hot water or hot air.

9.2.2   Evacuated Tube Collector

Evacuated tube collectors consist of a heat pipe inside a vacuum-sealed tube that uses liquid–vapor phase change material (PCM) in an evaporating–condensing cycle to transfer heat from the absorbed solar radiation to a circulating fluid at high efficiency. The absorber with a cylindrical or flat shape placed in the vacuum envelope collects solar radiation and evaporates the PCM. The vapor of the PCM then travels to the upper part of the heat pipe, in the sink region, where it condenses and releases its latent heat to a fluid circulating in the manifold. The condensed fluid returns to the solar collector and the process is repeated.

Therefore, several evacuated tube collectors can be connected to the same manifold to composed a whole module. Another possibility is to connect these evacuated tube collectors directly to a hot water storage tank. The reduction of convection and conduction losses due to the vacuum envelope result in better performance and in operating at higher temperatures than flat plate collectors. They also collect both direct and diffuse radiation. Their efficiency is good at low incidence angles too, which gives them an advantage over flat plate collectors in terms of daylong performance.

Figure 9.4 shows an example of an evacuated tube collector module integrated to the roof whereas Fig. 9.5 shows an example of evacuated tube collector connected directly to a hot water storage tank for hot water application.

Fig. 9.4 An evacuated tube collector module integrated to the roof (Creator: F.Mykieta, Source: link, License: CC BY-SA 4.0)

Fig. 9.5 Evacuated tube collector connected directly to a hot water storage tank (Creator: Rkraft, Source: link, License: CC BY-SA 3.0)

Fig. 9.6 Compound
parabolic collector

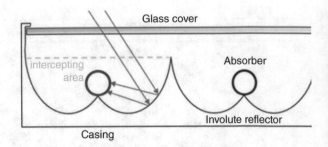

9.2.3   Compound Parabolic Collector

Compound parabolic collectors (Fig. 9.6) are part of the non-tracking solar collectors category. They are then stationary systems, but they are nevertheless a concentrating solar collector, which can be seen as an exception from the two main categories described before.

They consist of two wings of reflectors. Each wing belongs to a different parabola. One parabolic reflector helps to collect the incoming rays when the sun is on eastern side and other wing collects the incoming rays when the sun is on western side. Thus, these collectors allow sun rays over a wide range of angles to enter the concentrator mouth and reach the receiver through internal reflections.

Using concentrators and appropriate tracking, the solar thermal collectors can also be used for power generation application which has been discussed in subsequent section.

9.3   Application of Solar Thermal Energy for Power Generation

Power generation is usually made through a process where thermal energy from a heat source is converted into mechanical energy with a turbine and then converted again into electricity with a generator as shown in Fig. 9.7. This process commonly uses a thermodynamic cycle that runs the turbine between a gap of temperature. Higher the temperature, more efficient is the turbine, and higher the electricity generation.

This thermodynamic cycle circulates a working fluid (mostly steam) that brings heat from the source to the turbine before being released to the cold source. The heat from the source is transferred to the working fluid with a heat transfer fluid and is released to the cold source with a cooling fluid.

Several heat sources can be used in a power generation process. Some of them are from fossils resources like coal, gas or nuclear. And some others are from renewable resources like biomass and of course the Sun. About the cold sources, water or the ambient air is generally used.

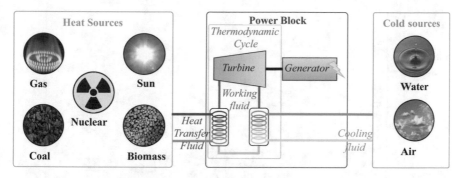

Fig. 9.7 Process of power generation through heat source

For power generation processes that convert solar energy into electricity, the amount of heat collected from the Sun needs to reach a high level to run steam turbines, far above those attainable by flat plate collectors. Thus, an optical device is used to concentrate solar radiation and reach much higher temperature. There are four main concentrating collectors that are mostly used in power generation processes which are as follows [2]:

– Parabolic trough collector
– Linear Fresnel reflector
– Parabolic dish reflector
– Heliostat field collector

These collectors are presented in Table 9.2 in function of its absorber type, whether it is fixed or mobile, with a line or a point-focus and whether it tracks the Sun on 1 axis or two axes.

The first of them is the Parabolic Trough Collector. It consists of parabolic shaped mirrors putting in line that concentrate the solar radiation onto a linear tubular absorber placed along the focal line of the collector, and where a fluid absorbs the heat to be used. This one-axis tracking solar collector is the most advanced of the solar thermal technologies and is the most mature because of considerable experience with the systems and the development of a small commercial industry to produce and market these systems. Parabolic Trough Collectors are built-in modules that are then connected together in order to form a solar field of collector.

This collector can be oriented in an East–West direction, tracking the sun from north to south, which enhances the collector performance during the middle of the day and during winter. It can also be oriented in a north–south direction, tracking the sun from east to west, which enhances the performance during the early and late hours of the day and during summer. The east–west field provides a more constant annual output, but over the year, a north–south trough field usually collects more energy than an east–west one.

Therefore, the choice of orientation usually depends on the application and whether more energy is needed during summer or winter.

Table 9.2 Solar concentrators for power generation

Tracking	Absorber	Mobile	Fixed
1-axis	*Line-focus*	**Parabolic Trough Collector (PTC)**	**Linear Fresnel Reflector (LFR)**
2-axis	*Point-focus*	**Parabolic Dish Reflector (PDR)**	**Central Receiver Collector**

Figures 9.8 and 9.9 show some examples of concentrating solar power plants using parabolic trough collectors at Mojave desert in California.

Linear Fresnel collector consists of an array of linear mirror strips that concentrate light onto a linear receiver. From another point of view, it can be seen as a parabolic trough concentrator that have been cut out into flat strips and then put closer to the ground. Thus, unlike the parabolic shaped mirror of a parabolic trough, individual flat strips reflectors are a way cheaper to manufacture and have less structural requirements. Nevertheless, the disadvantage of this technology is the shading that can occur between adjacent reflectors, that imply a smaller concentrating ratio than parabolic troughs.

Figures 9.10 and 9.11 show some examples of CSP plants using linear Fresnel collectors where Fig. 9.10 corresponds to the CSP Fresnel power station Puerto Errado 2 from Novatec Solar and Fig. 9.11 corresponds to a test loop at Sandia National Laboratories at Albuquerque in New Mexico.

Heliostat field collector is essentially used in tower CSP plants. It consists of multiple flat mirrors, or heliostats, of around 50–150 m^2 where each of them concentrates the direct solar radiation onto a central receiver at the top of the tower. The concentrated heat energy absorbed by the receiver is transferred to a circulating fluid, mostly a gas since it can reach extremely high temperature (above 2000 °C). The heat is then used to produce steam at high temperature and pressure.

Central receivers have several advantages:

– They collect solar energy optically and transfer it to a single receiver, thus minimizing the thermal energy transport requirements.
– They typically achieve concentration ratios of 300–1500 and so are highly efficient, both in collecting energy and in converting it to electricity.
– They are quite large (generally more than 10 MW) and thus benefit from economies of scale.

Here are some examples of CSP plants using heliostats and central receiver as shown in Figs. 9.12 and 9.13. Figure 9.12 corresponds to Themis, the first pilot of a solar power plant in the world, at Targassone in France and Fig. 9.13 corresponds to the Crescent Dune solar power plant of 110 MW in the Nevada desert of USA.

Parabolic dish reflector consists of a paraboloid shaped mirror that concentrate the solar radiation onto a point-focus absorber and tracks the sun in two axes. It is the most efficient of all collectors, with the highest concentration ratio. Nevertheless, its operation and its application are quite different from the other collectors presented previously. Indeed, they are much smaller than the other point-focus type absorber of the previous heliostat field collector, which imply that they are rather made to be modular. Although they can be assembled in the field, where they can each collect heat from solar radiation with a fluid circulating in the absorber, and then be transported through pipes to a central power conversion system, it raises design issues such as piping layout, pumping requirements and thermal losses, especially when the fluid temperature can be above 1500 °C. Thus, they are more made to be independent and self-sufficient by converting directly heat into electricity with its receiver unit located at the focal point. It is generally an engine-generator based on a Stirling thermodynamic cycle, better known as a Stirling engine, where the heat

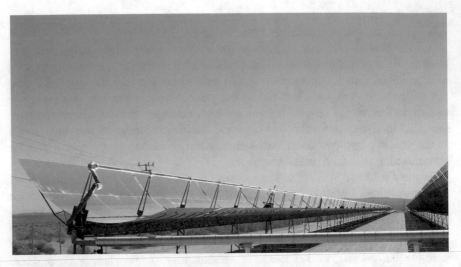

Fig. 9.8 Parabolic trough collector (Creator: Kjkolb, Source: link, License: CC BY-SA 3.0)

Fig. 9.9 Field consisting of several parabolic trough collectors (Creator: Alan Radecki Akradecki, Source: link, License: CC BY-SA 3.0)

Fig. 9.10 Fresnel power station Puerto Errado 2 (Creator: Novatec Solar, Source: link, License: CC BY-SA 3.0)

Fig. 9.11 Test loop consisting of linear Fresnel reflectors at Sandia National Laboratories (Creator: Sandia Labs, Randy Montoya, Source: link, License: CC BY-NC-ND 2.0)

Fig. 9.12 Pilot of a solar power plant consisting heliostats at Targassone in France (Creator: Unknown, Source: link, License: CC0)

Fig. 9.13 Crescent Dune solar power plant in the Nevada desert (Creator: Amble, Source: link, License: CC BY-SA 4.0)

Fig. 9.14 Parabolic dish reflector (Creator: The U.S. National Archives, Source: link, License: No known copyright restrictions)

Fig. 9.15 Solar field consisting of parabolic dish reflectors for power generation (Creator: United Sun Systems International Ltd, Source: link, License: CC BY-SA 4.0)

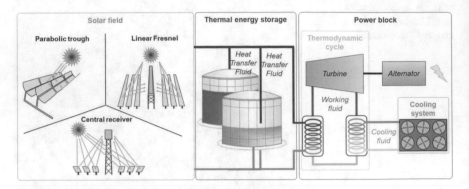

Fig. 9.16 Concentrating solar power plant integrated with thermal energy storage

source is concentrated solar radiation instead of crude oil or coal. Figures 9.14 and 9.15 show some examples of single and a field of parabolic Dish Reflectors with Stirling engine at the focal point.

9.4   Summary

To summarise, power generation processes that use concentrated solar energy as heat source are based on the same principle than those using other energy resources (like coal, gas, nuclear, etc), especially for the conversion of heat into electricity that is made in the power block. The only difference is that CSP plants, need to deal with the natural fluctuation of the resource (like the day/night cycle, clouds, etc). Thus, a thermal energy storage can be added in these processes to timely decouple the solar energy collection in the solar field, from the heat conversion in the power block as shown in Fig. 9.16. This allows for a smoother electricity production which facilitates its supply to the utility grid.

References

1. Duffie J.A., Beckman W.A., Solar Engineering of Thermal Processes- Fourth Edition, Wiley, New York (2013)
2. S.P. Sukhatme, J.K. Nayak, Solar Energy Principles of Thermal Collection and Storage (Third edition), The Tata McGraw Hill Education Private Limited, New Delhi (2009)

Index

Printed in the United States
by Baker & Taylor Publisher Services